Developing and Managing Innovation in a Fast Changing and Complex World

Stig Ottosson

Developing and Managing Innovation in a Fast Changing and Complex World

Benefiting from Dynamic Principles

Springer

Stig Ottosson
Department of Manufacturing and Civil Engineering
Norwegian University of Science and Technology—NTNU
Gjøvik
Norway

ISBN 978-3-030-06784-7 ISBN 978-3-319-94045-8 (eBook)
https://doi.org/10.1007/978-3-319-94045-8

Softcover re-print of the Hardcover 1st edition 2019

Printed on acid-free paper

This Springer imprint is published by the registered company Springer International Publishing AG part of Springer Nature
The registered company address is: Gewerbestrasse 11, 6330 Cham, Switzerland

Preface

It is difficult to know where is best to start treating such a complex topic as management of innovation and the development of innovations. However, based on my experience, I think it is good to start to give you some basic statements before we go into more details.

In short, innovations are new products (meaning goods and services) that are sold and used. Therefore, a new product or an invention is not an innovation but can be part of it. With sold is meant a wider view so that money does not need to be exchanged.

To coordinate all activities to develop an innovation is a complex management activity for which some parts can be planned and for which *creativity* and *improvisation* are needed when planning is not possible or rather meaningless to do in depth.

> *Creativity* means shakings things up, both inside ourselves and in the world around us, and constant reorganizing of both cognitive schemata and, to a greater or lesser extent, the domain of the creative person's activity. (Montuori 2003)

> *Improvisation* is thought of as making the best of things, while awaiting a return to the way things should be done. Improvisation is an exception, something we can "fall back on" when things don't go the way they should. (Montuori 2003)

Therefore, innovation management is different from management of mature businesses for which most things are well planned. Also, innovation management is place and situation dependent, meaning that the culture in the organization and the geographical area where the development takes place and the market where the initial marketing and sales takes place.

The more complex the leadership situation in a region or country is, the higher the demands to produce and market sustainable innovations and the more complex the innovation management as well as cost will be. Sweden, where I live and was brought up, is an example of a country with such a demanding management situation (see Fig. 1). Although Sweden's population is only equivalent to 0.13% of the total world population, it is the 33rd largest export economy in the world and the fourth most complex economy according to the Economic Complexity Index (ECI).

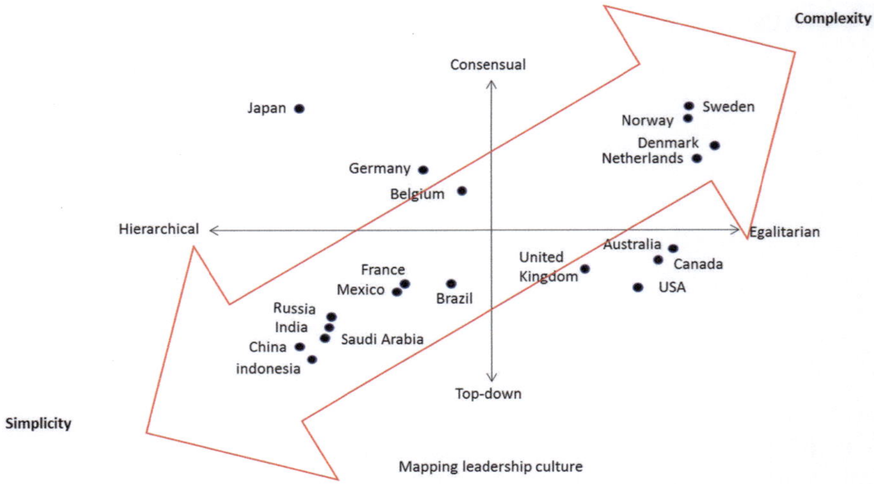

Fig. 1 Mapping leadership culture (Based on Meyer 2017)

Another example is that, although the Nordic countries, and especially Norway, have high labor costs with an average of 125 compared to the G7 countries of 108, they have managed to be successful on a global scale. "Therefore, there are important lessons to learn from high-cost countries that successfully compete in the global marketplace" (Holtskog et al. 2018, pp. 1 and 106).

An interesting example of what Fig. 1 illustrates is findings from a case study carried out at the vehicle R&D center China Europe Vehicle Technology, CEVT, at Lindholmen, Gothenburg (Holm and Zhang 2016). CEVT is a subsidiary of the Chinese Geely Holding group, which also is the owner of, for example, the vehicle manufacturers of Volvo Cars and Geely Auto.

Some conclusions in that report are that "Chinese and Swedish leaders have been inspired by different philosophies or models in their leadership, and they are usually connected to the cultural upbringing of these individuals. Every person's upbringing affects their way of being and functioning in work settings, which is something that the managers at CEVT try to take into account. This affects leadership as well, and can lead to misunderstanding but also development, especially when learning from each other."

My own experiences of innovation management and the development of innovations are mainly from the Nordic region of Europe, although, in that work, I have experienced the importance of cultural awareness when finding new markets for our innovations.

The leadership situation shown in Fig. 1 has strong connections to the cultural map (see Fig. 2). On the y-axis in that figure, traditional values emphasize the importance of religion, parent–child relationships, and authority. People who embrace these tend to reject divorce, abortion, euthanasia, and suicide. These societies also exhibit high levels of nationalism and national pride. On the x-axis,

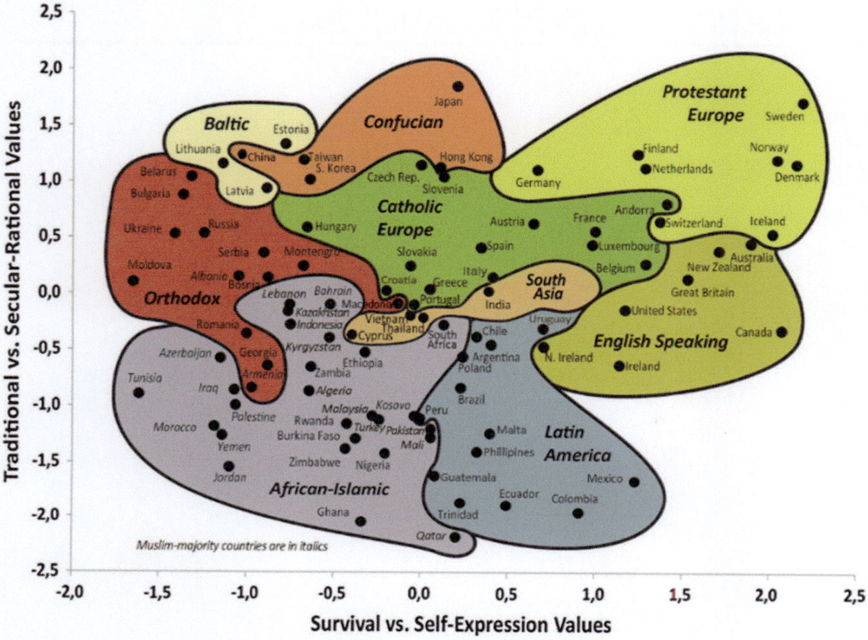

Fig. 2 The cultural map www.worldvaluessurvey.org/images/Culture_Map_2017_conclusive.png

survival values refer to economic and physical security and safety and are linked to low levels of trust and tolerance. On the other side, self-expression values give high priority to protecting the environment, promoting gender equality, and tolerating foreigners and homosexuals.

Note also that corruption that can be used to boost or reach positive results in business is not allowed or practiced in the Nordic countries (see Fig. 3).

Regarding situational dependencies for innovation management, there are, in principle, three main types of management situations: to develop incremental innovations managed by *project leaders* and managers, to develop more radical innovations managed by *entrepreneurs* and *intrapreneurs*, and to develop "survival innovations" managed by *renovateurs*. The three innovations appear within the *private sector*, the *public sector*, or in the *idealistic/nonprofit sector*. In the book, we will address all three types of management situations, the three leader types, and the three sector types.

Thus, we can distinguish between three main innovation types: *incremental innovations, radical innovations,* and *survival innovations.* The most common type of innovation is the *incremental innovation*, meaning to continuously develop products and services as well as business models and often also organizational settings until a disruption situation occurs (see Fig. 4) often initiated by a competitor in the market. Lean product development is one method to successfully

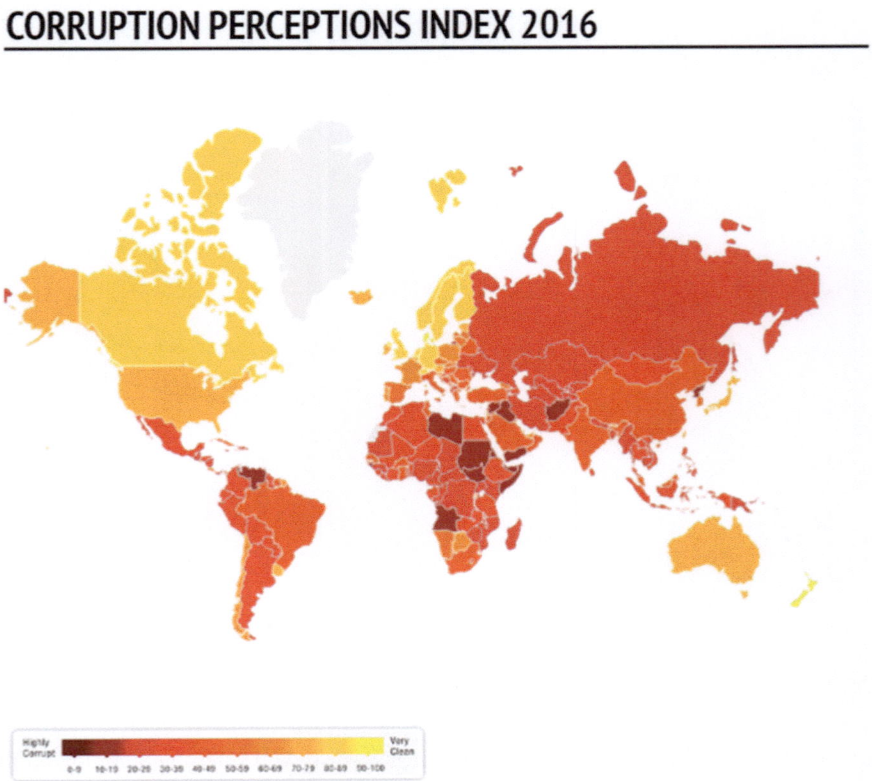

Fig. 3 The corruption map 2016 (www.transparency.org/news/feature/corruption_perceptions_index_20)

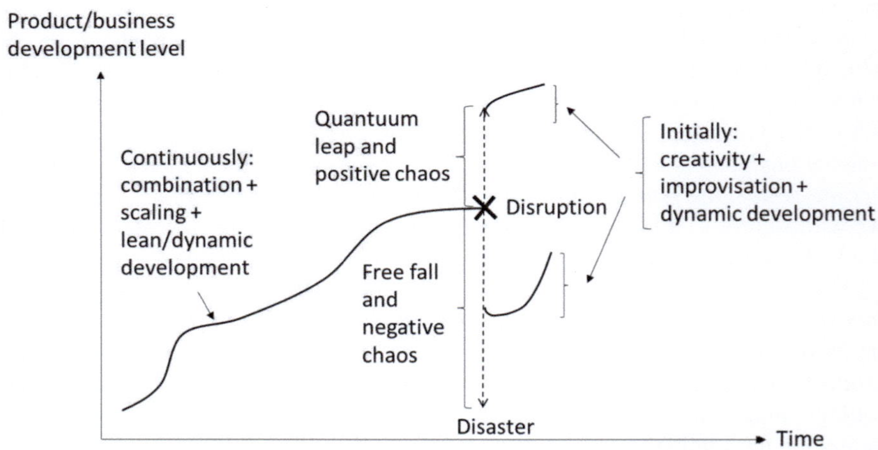

Fig. 4 Situational innovation (product and business) development

make continuous improvements of existing products and solutions. When a disruption situation occurs, there will be a free fall to stop or a new opportunity to develop based on new knowledge or a new solution that is sometimes is called a *quantum leap*. The free fall means a negative chaos, while a positive chaos and a euphoric feeling often occur when a completely new opportunity appears. Note that the curves in Fig. 4 are not smooth in real development projects, which we will come back to.

Radical innovations appear as sudden steps up in the development level, sometimes called quantum leaps (see Fig. 4). They are based on knowledge breakthroughs that are developed and introduced in the market. In that case the technical development content is often initially large, and that the organizational and business development comes successively after the introduction of the product on the market.

"*Survival innovations*" are needed for an organization when a radical change is needed for the business to survive. It can be when a sudden step down occurs (see Fig. 4), for example, because of new laws or environmental changes, or when a market drop occurs for the business. This occurs because competitors have introduced new technology and/or marketing and sales principles. In that case, the technical development content initially is often small and that the organizational and business development is dominant, striving to improve the situation.

As the most complex innovation management situation seems to be how to bring up a new hardware commercial product, the book deals with the management and development of such an innovation to a great extent. However, a software commercial product or a nonprofit innovation contains many parts of a commercial hardware product, so the book should also be valuable if your mission is to build up such an innovation in any sector or any size of organization. The conclusions are hard drawn, although, in reality, it is possible to find exceptions to the rule.

Only with technological innovations can climate and environmental challenges be overcome and welfare increased. A remark, which might be controversial, is that by avoiding purchasing new products and repairing old ones, less money can be used for the development, making our future more sustainable (Westergård 2016). Thus, a sustainability paradox is that the more we consume, the more sustainable the society will be as more money for development will be available. However, the consumption should be responsible and it is important that the politicians set increasingly hard demands on the products that are developed and sold. One of my ambitions with this book is to give useful knowledge in how to develop sustainable products and innovations.

Another ambition is to give useful rules of thumb based on a dynamic mind-setting to treat different situations. Therefore, I strongly urge you to study and reflect on what is described in Appendix A.

Each chapter is written to stand on its own feet, which is why some repetitions appear in the chapters.

References to work older than 1995 are not, in general, given in the book as useful knowledge from earlier work can be seen as having been settled in general knowledge. When no references are given, references are to my own experiences and to other peoples' experiences that I have got access in different ways.

Gjøvik, Norway Stig Ottosson

Acknowledgements

Dr. Alan Turing—Father of the Modern Computer, realized that: "— a prerequisite to find out something new is that we have the ability to make mistakes, to go beyond the framework. The one who always thinks correctly creates nothing. The one who only follows the current opinions or programming, and to talk machine language, does not come up with anything truly innovative and cannot even be considered intelligent in a real sense" (Lagercrantz 2016, p. 285). He also is claimed to have said that "The mystery is always bigger than the solution of the mystery" (Lagercrantz 2016, p. 356), which has as a consequence that "The solutions are not perceived as remarkable when you know the answers. You do not understand how difficult it was to find them until you have an answer key." (Lagercrantz 2016, p. 365)

Rendering and connecting to that, this book is the result of a search for methods and explanations since 1979 (see Fig. 5) of why a dynamic approach is more effective than the traditional Newtonian approaches for which disorder, chaos, and noise are not taken into account (e.g., Montuori 2003). My findings are based on countless mistakes, some successes, my questioning of different perceptions, extensive literature studies, countless conversations, and discussions, as well as amounts of creative incidents and tests of them together with combining bits and pieces to form a useful entity.

Although the writing of the book was a solitary activity, without the uncountable inputs from uncountable numbers of people it would not have been possible to write it. All the students I have had in the Nordic countries, Germany, and some other countries have helped me to reflect on the topic and to see more clearly and understand the complexity of creating sustainable innovations independent of whether it is within the private sector, the public sector, or the idealistic sector. However, especially I would like to thank Prof. Evastina Björk, Professor Emeritus Sandor Vajna, and Mr. Sri Kolla for our talks and discussions and their valuable

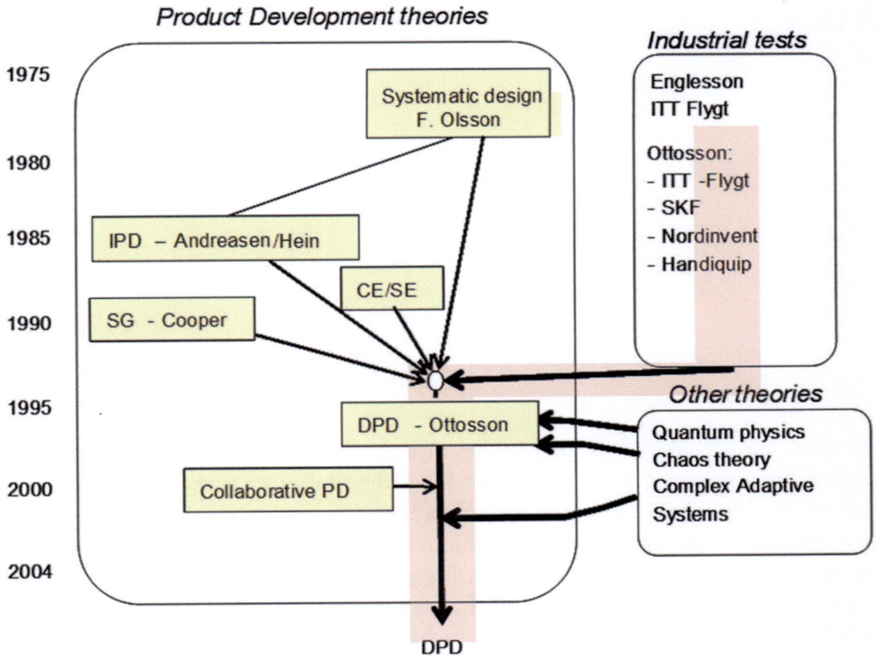

Fig. 5 The development of dynamic principles and methods from 1979 onward by combining and adding development principles based on industrial tests and theories (Ottosson 2006)

inputs and honest critical stance when I have made conclusions without solid ground to stand on. Not the least that has taught me how important it is to verbalize ideas to find out the weaknesses.

I also want to thank Mrs. Carol de Groot who has helped me with the language controls as well as pointing out unclear writing in the original text.

April 2018 Stig Ottosson

Contents

Acronyms

AR	Action Research and Augmented Reality
ASD	Agile Software Development
B2B	Business-to-business
B2C	Business to consumer
BAD	Brain-Aided Design
C&E	Cause and Effect
CAD	Computer-Aided Design
CE	Concurrent Engineering
CEO	Chief Executive Office
CI	Cultural Identity
CMS	Content Management System
CPC	Cost per click
CPM	Critical Path Method
CRM	Customer relationship management
Dfa	Design for all
DFMA	Design for Manufacture and Assembly
DfU	Design for Usability
DfX/DFX	Design for X
DMU	Digital Mock-Up
DPD	Dynamic Product Development
EDM	Engineering Design Management
F2F	Face-to-Face
FAQ	Frequently Asked Questions
FFE	Fuzzy Front End
FFF	Freeform Fabrication
FTA	Failure Tree Analyzes
Glocal	Global and local
HDM	Head-Mounted Displays
HR	Human Relations
IAR	Insider Action Research

ICT	Information and Communication Technology
IoT	Internet of Things
IP	Intellectual Properties
IPD	Integrated Product Development
ISO	International Organization for Standardization
LABDA	Look, Ask, Model, Discuss, and Act
LCA	Life Cycle Analyzes
LPD	Lean Product Development
M2M	Machine-to-machine
MAD	Model-Aided Design
MBWA	Management by walking around
MLM	Multilevel Marketing
NPD	New Product Development
PAD	Pencil-Aided Design
PAR	Participation Action Research
PCT	Performance, Cost, and Time
PD	Product Development
PDM	Product Data Management
PERT	Program Evaluation and Review Technique
PLC	Product Life Cycle
PMU	Physical Mock-Up
PPC	Pay per click
PR	Public Relations
Q	Quality
QFD	Quality Function Deployment
QR code	Quick Response
R&D	Research and Development
SEO	Search Engines Optimization
SG	Stage-Gate
SME	Small- and medium-sized enterprise
Specs	Specifications
SWOT	Strengths, Weaknesses, Opportunities, and Threats
TPS	Toyota Production System
TRIZ	Theory of Inventive Problem Solving
UCD	User-Centered Design
UD	Universal Design
USP	Unique Selling Points
VC	Venture Capital
VR	Virtual Reality
www	World Wide Web

List of Figures

List of Tables

Chapter 1
Introduction

1.1 What Is Meant by Products and Innovations?

A product generally is seen as something we can use, but what is innovation or an innovation?

The term "innovation" seems to derive from the Latin "novus", which means new or young or novel. Unfortunately, there is no single accepted definition of the term "innovation". Historically, innovation was defined as the introduction of new elements or a new combination of old elements in industrial organizations (Schumpeter 1934). Thus, the actual new idea or invention but also the realization of them was its focus. Much later, Kanter (1983) defined innovation as the process of bringing any new, problem-solving idea into use.

If today we think of "innovation" as a substantive "innovation project", will have as our mission to do all activities on the micro level in an organization, develop, market and "sell" a new product and/or service with the aim that it shall be taken into use or be consumed. This book has that view and a short definition of innovation covering all sectors of the society and related terms therefore might be:

> Innovations are new products and/or services that have been "sold" and taken into use in at least one local market.

This is close to the definition in the Oslo Manual (OECD 1997), where it is formulated as: "An 'innovation' is defined as the implementation of a new or significantly improved product (good or service), and/or a new organizational method in business practices or workplace organization Basically three different start situations for the development are at hand: adding to a continuous situation (linear or un-linear), taking care of new knowledge from science or an invention (creating a quantum leap innovation), or breaking a free fall from an unplanned disruption through a new start (change management)."

S. Ottosson, *Developing and Managing Innovation in a Fast Changing and Complex World*, https://doi.org/10.1007/978-3-319-94045-8_1

Connected to that definition are some essential terms:

An *innovation process,* done as an innovation project, will contain all of the stages from idea generation, development and commercialization, to an implemented solution on the market.

The leader of a development project can be called the *project leader* or *manager*, which also counts for incremental innovation projects. The leader of an innovation project is normally called *entrepreneur.* The leader of an innovation process in an existing company or organization is sometimes called *intrapreneur* (Bjerke 2013, p. 90). The intrapreneur normally has limitations in the work that an entrepreneur does not, for example, having to hire people and to sell the products through other channels than their own. The leader for a survival and/or change management project has been termed *renovateur* (Bjerke 2013, p. 90). Generally, the renovateur has a short time to act and lives under some psychological pressure because of the negative chaos situation in which she/he acts.

In general, innovation is necessary to bridge the gap between the present situation and, for example, to master critical new challenges caused by greater international competition, more demanding citizens, a growing elderly population, and environmental challenges.

- "Innovation is as essential to a successful modern economy as water is to life", according to EU Research Commissioner Maire Geoghegan-Quinn (2013).
- "Innovation and technology drive everything from healthcare to education to election wins" (Hoque 2013).
- "Today the public sector and the public services it provides are confronted by increasing demands, such as how to manage the complex issues of demographic and climate change, or how to respond to users' diverse demands" (www.oecd.org 2013).

There are many ways to classify innovations. Based on the start of their development, innovations can be seen as technological innovations sometimes called technology push innovations in the private sector) or behavioral innovations (sometimes called market pull innovations in the private sector).

Based on their development start innovations can be technological or behavioral. Some examples of modern innovations are:

Technological

- Ingenious medical devices
- Hybrid cars
- Smartphones
- Search engines

Behavioral

- Starbucks has transformed the experience of drinking coffee
- Skype has opened up voice communications
- Micro credits were developed to support entrepreneurship, to alleviate poverty, to empower women, and to uplift entire communities
- Distance learning.

Especially, to develop a sustainable society, innovations are needed to ensure that nature shall not be subject to the systematic increase of (www.thenaturalstep.org):

- concentrations of substances from the earth's crust (such as fossil CO_2 and heavy metals),
- concentrations of substances produced by society (such as antibiotics and endocrine disruptors),
- degradation by physical means (such as deforestation and draining of groundwater tables),
- structural obstacles to people's health, influence, competence, impartiality and meaning.
- A sustainable society is dependent on sustainable solutions. In this book we have used the following definition:

A *sustainable solution* is a solution that has been developed to be a long lasting, environmentally responsible solution for the provider (the business), the society and also the users.

Figure 1.1 shows the environmental impact from the "cradle" to the "grave" of a new product. The project leader of a new product development process and the product development team will strongly influence the environmental impact during the whole Product Life Cycle (PLC) of the product and the supplementary products that the primary product may need to function as intended. For many products, the use phase contributes significantly to the total environmental impact of the product's life cycle (Wever et al. 2008) although the production phase can have an environmental impact and it takes long time before the products can be used. This has been much discussed when considering the environmental effect of electrical cars compared to diesel or petrol driven cars.

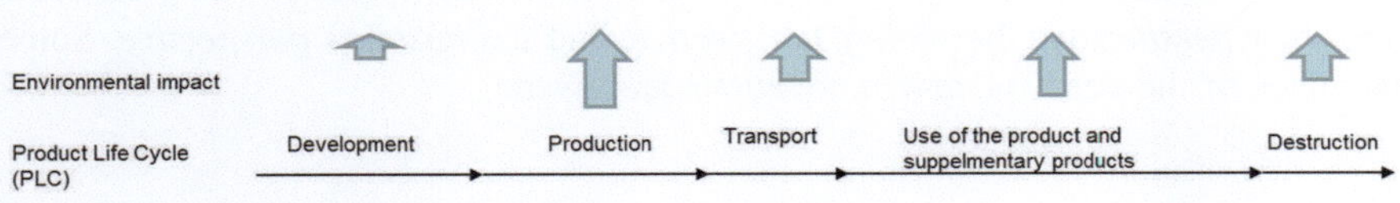

Fig. 1.1 Actions to develop a sustainable product solution (Ottosson et al. 2016)

As we cannot influence the past, it is of utmost importance that when developing new products, we do our best to make them sustainable. This is because as long as they are used, they will have an impact on the environment and the society we are living in. The product developers, therefore, have a great responsibility for our lives and for the lives of coming generations.

From the product developers, modern product development requires technical and non-technical skills, adequate practice, an emphasis on lifelong learning, monitoring of outcomes and a supportive institutional service framework. These requirements need to be combined with a number of personal and professional values including integrity, professionalism and compassionate, user-centered understanding (Agha et al. 2015).

Non-technical skills can be defined as a collective term describing the skills and behaviors encompassing situational awareness, decision making, communication, teamwork and leadership (Ind. Psy. 2012). Others have defined non-technical skills along three dimensions, including; the interpersonal (e.g., communication, teamwork), cognitive (e.g., decision-making, situational awareness) and personal resource skills (e.g., coping with stress and fatigue) (Flin et al. 2008).

1.2 Innovations in Different Sectors of Our Society

According to conventional understanding, innovations are only done in the private sector (Mulgan 2007). However, in reality, many have been and are being developed in the public sector and in the idealistic (non-commercial) sector. For example, from the public sector, for example, we have the Internet (CERN 1989), the World Wide Web (DARPA), and new teeth, repaired teeth and prostheses using titanium (Gothenburg University). In the idealistic sector, different open-source solutions have been and are frequently developed. When the new solutions in these sectors mature, they often "migrate" into the private sector to become commercial products and this might be why we see innovations as something emerging from the private sector.

Thus, innovations are and must be created and developed in all three sectors, although the aim of the work differs. For the private sector, the aim is mainly to create a sustainable profit. For the public sector, the main aim is to provide better service to the people in the society. For the non-commercial sector, a better world—on the local or global scale—is often the aim of innovative activities. Figure 1.2 shows these aims for the three sectors.

The need for new innovations can be seen from three points of view: the *user/consumer perspective*, the *society perspective*, and the *business perspective*. Some examples of the need for new innovations are given.

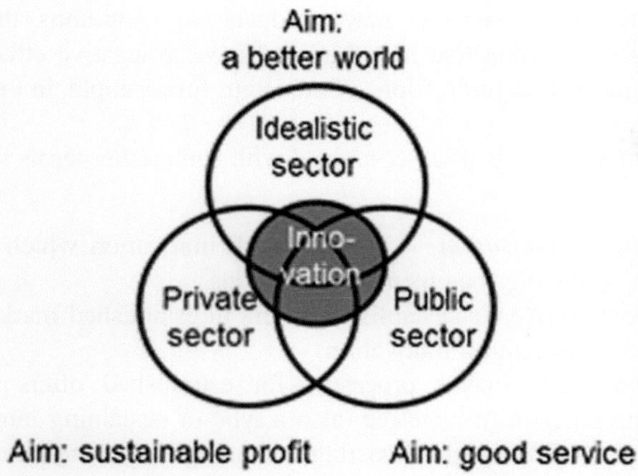

Fig. 1.2 Innovative work in different sectors has different aims

User/consumer perspective:

- to provide needed or wanted new solutions and new products and services
- to satisfy an environmental awareness
- to support development for a better world

Society perspective:

- for economic growth
- for the progression of human well-being
- for creating a sustainable world

Business perspective:

- for business survival
- for better economic returns
- to take advantage of opportunities
- to improve disappointing performance
- to provide higher business revenues
- to replace cost-cutting, which causes anorectic organizations
- for competitive advantages in hiring professional people
- for competitive market advantages

Some additional views of innovation, which deepen the discussion include:

- "the introduction of a new idea, method, or device" (White and Glickman 2007, p. 97)
- "change that creates a new dimension of performance" (Hesselbein et al. 2002)
- "a creative act or solution that results in a quantifiable gain" (Rosenfeld 2006)
- "next practice" (Engholm Jensen 2008).

Unfortunately, there are also new products and solutions that have been implemented on the market that have had and have a negative effect on mankind and the environment. Bad innovations often occur, for example, in the black market and military sectors.

The term "innovation" is often combined with some other terms such as (Moore 2004):

- *radical or basic innovation*—a breakthrough innovation which creates a new platform for following incremental innovations
- *product innovation*—takes established offers in established markets to the next level (a type of sustaining innovation)
- *process innovation*—makes processes for established offers in established markets more efficient or effective (also a type of sustaining innovation)
- *application innovation*—takes existing technologies into new markets to serve new purposes
- *experiential innovation*—makes cosmetic/surface modifications of established products or processes that improve customers' experience
- *marketing innovation*—improves customer touching processes e.g. by marketing communications or consumer transactions
- *business model innovation*—reframes an established value proposition to the customer or a company's established role in the value chain or both
- *structural innovation*—capitalizes on disruption to restructure industrial relationships
- *public sector innovation*—new or significantly improved ways of doing things, either within the structure of the public sector itself, or in the way in which public services are provided (OECD).

Within the public and idealistic (non-commercial) domains, the "customers" get access to the new solutions, often without paying for them in monetary terms. Examples of such innovations are education innovations, administrative innovations, political innovations, nursing innovations, or cultural innovations.

1.3 Three Lean Development and Innovation Development

Quite often, the term "Lean" is used in all the three sectors as shown in Fig. 1.2. Therefore, some words must be said about that term. *Lean development* has the aim to limit or take away all types of waste, such as time delays, unnecessary spending of money spending or unnecessary use of materials. As innovation development is a learning process, a lot of waste often occurs before a successful solution or handling is reached. Therefore, innovation and Lean are contradictory to each other until a functional solution has been developed. However, when a functional solution has

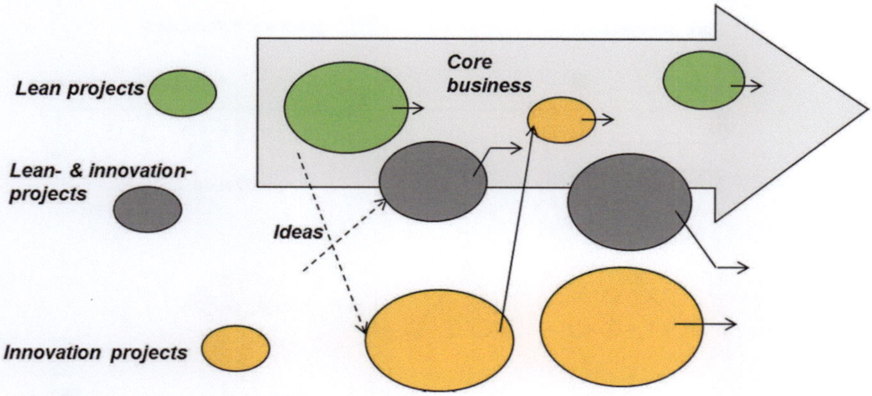

Fig. 1.3 To improve the core business, Lean projects are used. To develop new innovations, innovation projects are used

been reached, Lean development is needed for the innovation to be further developed to be a sustainable (economic) business for the organization.

Lean projects are normally done in the core business of an organization. Incremental innovation projects are generally done in the vicinity of the core business, with the aim to either implement them in the core business when they have reached a sustainable level or to develop them further outside the core business. For radical innovation projects, it often is better to do the development outside the core business as there are few synergy effects with the core business—they simply differ too much from the core business to live well in environment of the core business. Figure 1.3 shows the three possibilities of development projects.

Before an innovation has been developed to be rather mature, it is extremely vulnerable and needs to be taken care of in a very sensible way, which means providing shelter, care, nutrition, and own localities. Figure 1.4 shows a fruitful situation of how a larger organization can take care of its Lean and innovation activities side by side.

More detail follows on how an innovation organization is best set up, but as an introduction, Fig. 1.5 shows how the different innovation projects in Fig. 1.4 can be designed.

1.4 What Is the Difference Between Research and Development?

In this book, a 'theory' is seen as a system of assumptions, principles, and relationships posited to explain a specified set of phenomena. A 'method' is a way, technique, or process for doing something. Empirical methods use experimentation,

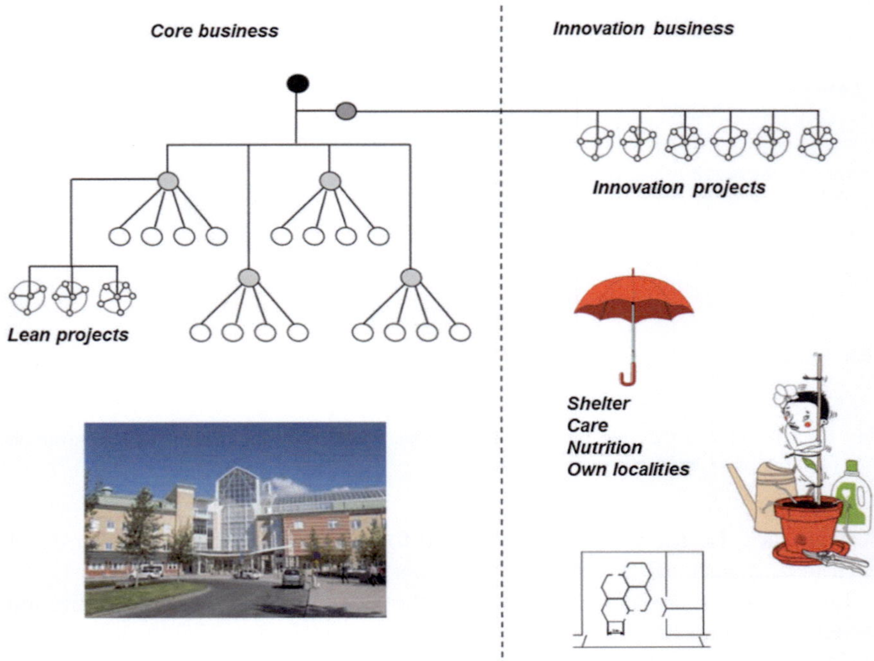

Fig. 1.4 Lean projects can be handled in the core business while Innovation projects are best taken care of outside the core business

observation and experience rather than logic and theory to extend knowledge. A methodology is often a whole set of methods developed according to a philosophical theory about how best to research and learn about natural or social phenomena. *Methodology deals with the general principles or axioms of the generation of new knowledge. It refers to the rationale and the philosophical assumptions that underlie any natural, social or human science study, whether articulated or not. Simply put, methodology refers to how logic, reality, values, and what counts as knowledge has informed research* (McGregor and Murname 2010).

What then is the difference between research and development? Research is planned searching or critical investigation aimed at the discovery of new knowledge, potentially leading to new products, processes or techniques. Development can be seen as interpreting research findings or other knowledge into a plan or design of new products, processes or techniques. Development excludes changes to current products, services, processes or techniques made as part of or as a result of a company's day-to-day operations.

Both research and development can be activities of laboratory research aimed at the discovery of new knowledge, the formation of concepts and the design for

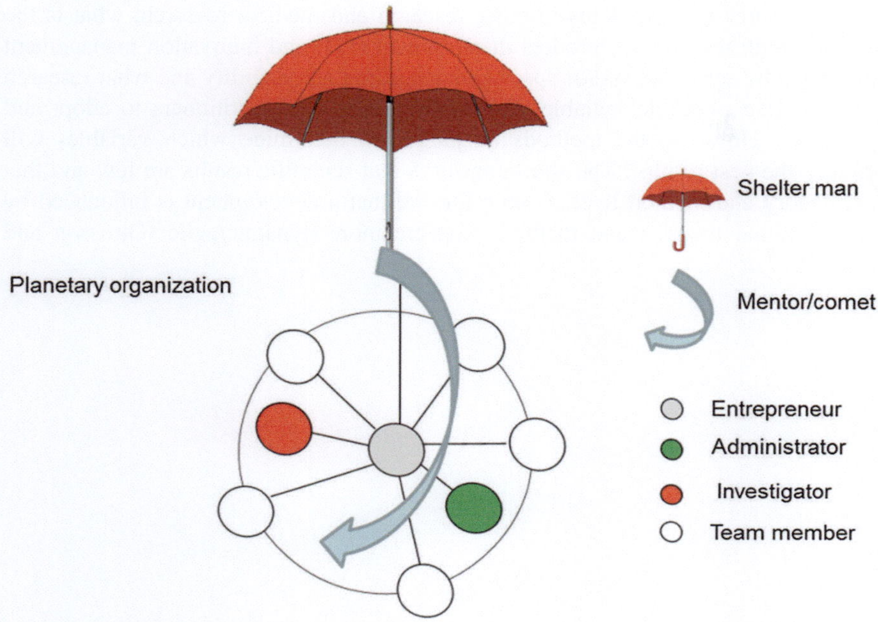

Fig. 1.5 The organization of an innovation project

potential product or process alternatives. Research and development activities also include the search for ways to apply new research, and the testing needed to evaluate potential product or process options. Thus, research and development can be the same thing.

Is a researcher also a scientist or is a scientist a researcher? A general perception is that any person involved in scientific research, making the activity her/his vocation, can be labelled as a researcher. A scientist is often seen as a person who has been trained and is an expert in one or more areas of science, and performs experiments to conduct research in a scientific manner. A scientist indulges in research to strive for a better, improved, and a healthier world for us. She/he also endeavors to enrich our understanding and knowledge about nature and other subjects, based upon science. Scientists are different than engineers, who spend their time grappling to come up with better products and appliances. However, a scientist can also be an engineer, for example, when developing equipment needed for research.

The general perception of scientists is that they should be objective seekers of truth, for which repeatability or reproducibility is important. That is, other scientists should be able to achieve the same or similar results using the same set up of experiments under the same conditions. However, this is not easy. Even in basic research and medical research, it has proven difficult to obtain the same results in repeated tests.

If there are such problems in basic research and medical research, what is the situation with research on product development (PD) and innovation management performed by scientists and/or specialists regarding repeatability and what research methods give plausible, reliable, and useful results for practitioners to adopt and adhere to? How can the methods be judged to determine which variables will produce the best results? The short answer is that scientific results are few and that each theory and method lives its own life but that a development is influenced by the traditional theories and methods that are more dynamic/agile (Ottosson and Kolla 2017).

Chapter 2
Innovation Theory

2.1 Is Innovation a Buzz Word?

"Innovation" has almost become a buzzword and there is no one single definition of its meaning. In general, it is a positively loaded term that brings hope in difficult times for the actors in the private sector, the public sector, the idealistic (non-profit) sector as well as in whole economies.

However, how to create successful innovations is seldom explained. Even more unclear is how to develop "sustainable innovations" although we might have an intuitive feeling of that expression as developing something good for the society and its inhabitants.

Researchers and scientists seem to concentrate their studies on innovation systems and innovation clusters on a macroscopic level. Their recommendations on how to develop innovations and how to use of different techniques and tools are often vague for practitioners, as, for example "creativity is a driver of innovation and a key factor for the development of personal, occupational, entrepreneurial, and social competences and the well-being of all individuals in society." (http://create2009.europa.eu/index_en.html), or "[agility is the capability to react, and adapt to expected and unexpected changes within a dynamic environment constantly and quickly; and to use those changes (if possible) as an advantage" (Böhmer et al. 2015). That it often takes many years to judge if an innovation project will be economically successful is another matter altogether and rarely discussed.

When methods and tools are not focused, there is, in general, a political focus on research and development (R&D) but not on the marketing and sales of the products. The result is what is sometimes wrongly referred to as a paradox because all the investments made in R&D result in remarkably few innovations.

The aim of this chapter is to clarify the terms and conditions for the successful development of innovations, independent of whether they are carried out in the private sector, the public sector, or the nonprofit sector.

S. Ottosson, *Developing and Managing Innovation in a Fast Changing and Complex World*, https://doi.org/10.1007/978-3-319-94045-8_2

2.2 Different Views on Innovation?

The reason to develop a new solution is that, after further work, it can also result in an innovation. It is often based on the desire to take care either of a new insight, for example, from a debate, from advertisements, or from own reflections. The desire can also be based on taking care of new knowledge, which means using research results, spin off solutions, user solutions or inventions. A third possibility is to take care of challenges by developing new solutions (see Fig. 2.1).

A *spin off solution* is a new solution derived from an existing solution.

User solutions are done by end users (Henkel and von Hippel 2012) and lead users (von Hippel 2005). Such *end users* develop solutions to their own needs without having their sale in mind. Often the solutions can—after refinements, for example, by a producer or service provider—be innovations. *Lead users* are experts in their fields who also develop solutions for their own use but who think their solutions can be of use for others. One example is when a sportsman develops new equipment which results in her/his better performance. Another is when a surgeon develops a new instrument to be better able to do her/his work. A third is programmers making new software when they feel the commercial products are not good enough. Generally, if lead users develop new solutions, the solutions reached are often of high quality and usability compared to when the solutions are developed in a traditional, industrial way.

Inventions are solutions that can be patented, meaning that they can get market protection defined by international laws. They are mostly done by end users, lead users and private or professional inventors.

To find existing knowledge and solutions that can be used as input for the development of concepts sounds easy. In reality, it can be time and money consuming. Often, unsearched information comes from unplanned meetings and unsearched

Fig. 2.1 There can be three different initiators behind the desire to develop a new innovation

information from magazines that someone happens to browse, giving the igniting spark for further thoughts. Some people will give credit to fate or luck when such things happen, but, in fact, the reason is more likely to be a question of being open-minded and taking care of a feeling or the smell of business opportunities.

Also, finding needs, wants, and wishes that can be transformed into opportunities can be difficult as they are often diffuse and are hidden in other circumstances. Therefore, they need to be drawn out into the light and brushed off to be visible and interesting.

The opportunities to find and take care of new and future possibilities are much increased by having a wide contact net from which information will come. Also, the possibilities will be increased by taking part in conferences, by listening to speeches, by mingling with people, by watching TV programs, by reading popular science magazines, and by using idea search engines.

Projects, in general, are set up to make something unique, and they set cost limits and completion dates in advance. In general, a traditional product development project is based on a market *need.* Innovation projects differ from such traditional projects in that they often form demands from trial and error, they have no clear finishing dates or rolling cost limits and they have income from the sales of the new products. The initiation of an innovation project is often a *want* or a *wish* for a new product or solution.

When there is a big leap in the technical content compared to the existing solutions on the market, a basic innovation—or revolutionary innovation—is sometime called a "quantum leap innovation" (Hamm 2007). Figure 2.2 shows the principle of a quantum leap. A typical quantum leap or radical innovation was the electronic calculator when it substituted the mechanical calculator. From the first all-electronic desktop calculator in the early 1960s through the transition of calculators to commodity status in the late 1970s, electronic calculators served as an important target for design engineering efforts, resulting in the acceleration of technology during a critical time in the electronics industry and bringing a number of other innovations. Still another quantum leap was when Apple introduced the iPhone 2007 with an advanced calculator build in.

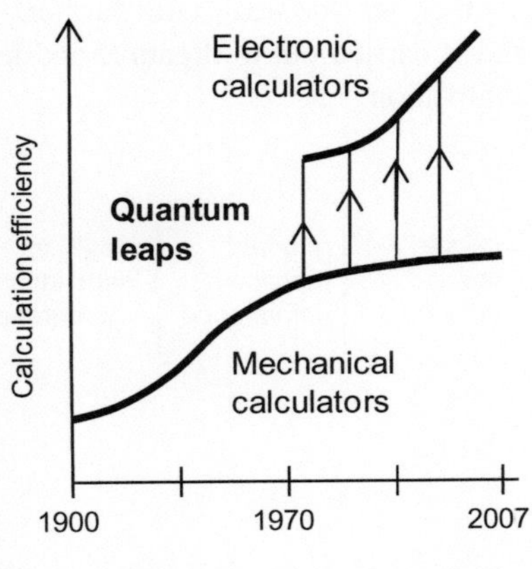

Fig. 2.2 A quantum leap means jumping from one level to another without first starting at a lower level

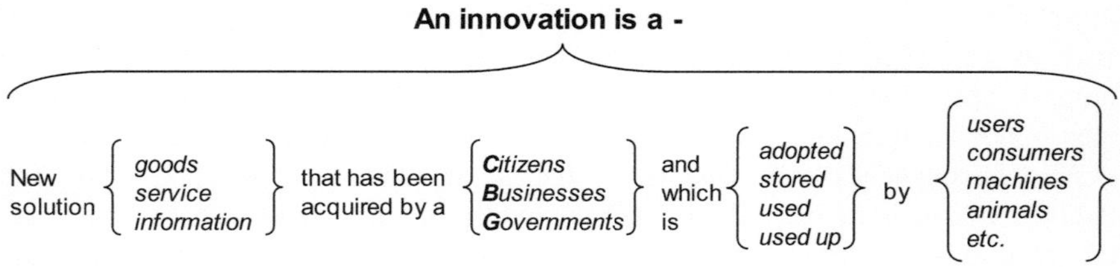

Fig. 2.3 A wider definition of the demands on an innovation

Note that as earlier explained, a new product that has not been bought or adopted by more than one user is not seen as a new innovation. A new product that only has been acquired and not been used is also not seen as a product innovation. "Acquiring" is seen here in a wider perspective than just an immediate payment of money.

Figure 2.3 shows a complete picture of what an innovation is. In the figure, "adoption" means that the product is stored, used, or used up/consumed.

Although it may sound strange, end users can be (human) users or consumers, animals, or machines. To make sure the products under development get to be "good enough", the product developers need to collect as much correct information as possible about their use, both before and during the entire development process.

A buyer's/customer's decision to acquire a product is dependent on many things, such as which buyer category the customer belongs to, the customer's present and/or future need of the product, the solution(s), the customer's emotional value of the product, the total price of the product during the time it is in the customer's possession and use, the financial situation of the customer, the estimated further sales price, and the positive or negative value of it at the time of its disposal.

Seen from the seller's side, she/he sells the product to a buyer (see Fig. 2.4). The seller can be a citizen, someone representing a business or a government. In e-business, citizens are abbreviated C, business B, and Government G. The whole chain from seller to a customer is then shortened to, for example, B2C. Sales are supported by marketing, which will be dealt with further.

Innovation management is carried out to organize and lead all activities in from an idea to a successful innovation.

Fig. 2.4 The sales chain

2.3 Diffusion of Innovation

Diffusion of innovation is a theory of how, why, and at what rate new ideas and technology spread through cultures. *Diffusion* refers to how an innovation has spread within a group, community or country. *Adoption* is more at the individual level (whether or not someone has used the innovation).

Everett Rogers introduced this in his 1962 book, Diffusion of Innovations, writing that "Diffusion is the process by which an innovation is communicated through certain channels over time among the members of a social system." Rogers theorized that innovations would spread through society in a pattern that looks like an S-curve (see Fig. 2.5). Pioneer buyers select the technology first, followed by the majority, until a technology or innovation is common. The effect will be a Bell curve as shown on the left of the Figure.

Thus, the diffusion of an innovation is the process in which it is communicated through certain channels over time among the members of a social system (Rogers 2003, p. 5). It is a special type of communication, in that the messages are concerned with new ideas. The communication is a process of convergence (or divergence) as two or more individuals exchange information in order to move toward each other (or apart) in the meanings that they give to certain events.

Diffusion research concentrates on the conditions that increase or decrease the likelihood that a new idea, product, or practice will be adopted by members of a given culture and people's attitude toward a new technology. Such research has shown that innovation diffusion is a process that occurs through five stages (Rogers 2003, p. 20):

- Knowledge of an innovation
- Persuasion (takes place when an individual forms a favorable or unfavorable attitude towards the innovation)
- Decision
- Implementation (when it is put in use)
- Confirmation

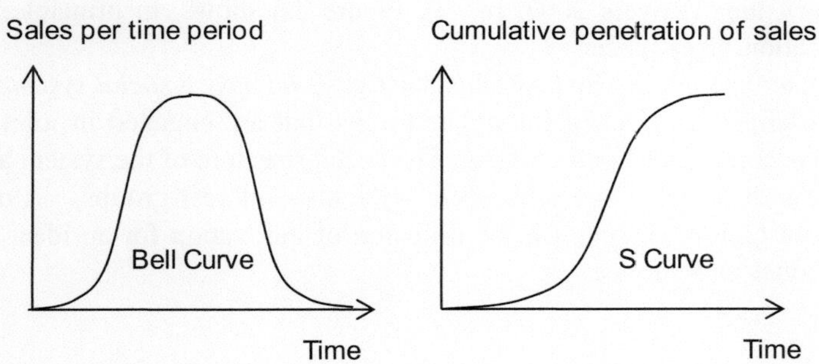

Fig. 2.5 The adoption curve becomes an S-curve when cumulative adoption is used

Thus, the buying process for a customer of a new product and/or service is the process through which an individual (or other decision-making unit) passes from knowledge of it, to the forming of an attitude towards the innovation, to a decision to adopt or reject, to implementation and use of the new idea, and to confirmation of this decision.

The buying process period for the first customer is the minimum length of time required to pass through the three steps, from knowledge to decision. Some people require many years to make the decision, while others move rapidly from knowledge to decision. For organizations, communities, and other types of adopting units, the decision process is often complicated as a number of individuals are involved in the process and have, for example, different political dimensions, or their own agendas.

2.4 Communication Channels

Diffusion of innovation is strongly influenced by the communication between people and how distorted the message becomes. *Communication* is a process in which participants create and share information with one another in order to reach a mutual understanding. When the participants have a goal to reach—a mutual understanding—the communication is done as a *dialogue*. Communication that does not have the goal of reaching a mutual decision but is intended to clarify the differences in opinions is done as a *discussion*. However, the result of a discussion can be, and often is, a new standpoint. A communication channel is the means by which messages get from one individual (the sender) to another (the receiver).

How the information is received is dependent on which channel is used, what kind of disturbances will influence the communication, and how well and fast the feedback is from the receiver to the sender. Mass media channels and the Internet are effective in spreading knowledge of innovation to a larger community, whereas inter-personal channels are effective in forming and changing attitudes toward to new idea on an individual level, and thus in influencing the decision to adopt or reject a new idea (Rogers 2003, p. 35). Figure 2.6 shows, in principle, how the communication takes place.

When many people are involved in an activity, we have a *social system*. A social system is defined as a set of interrelated units that are engaged in joint problem solving to accomplish a common goal. The social structure of the system affects the innovation's diffusion. Especially social systems—interest groups, set up on the Internet, can rapidly accomplish the diffusion of innovation for an idea, an initiative, a product or a service.

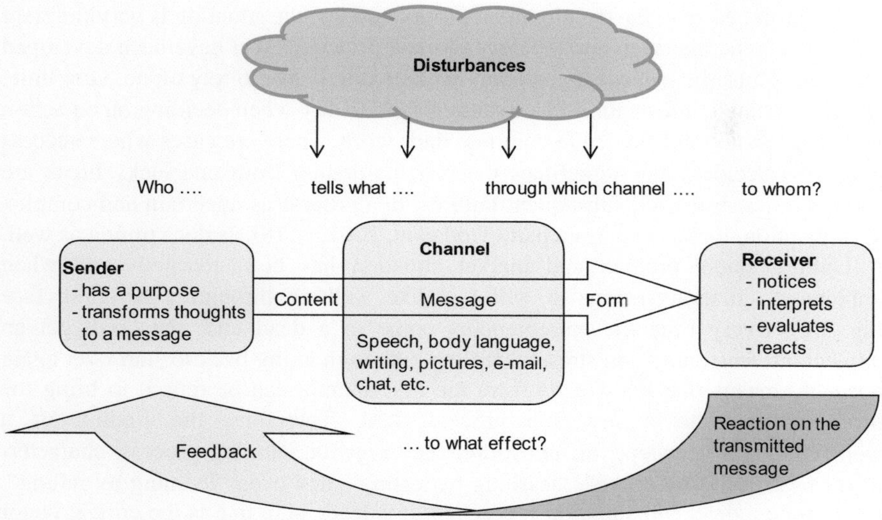

Fig. 2.6 The two-way communication loop

2.5 Innovation as a Learning Process

For the development of innovations, the classic theories tell us "Do it right the first time" (e.g., Cooper 2001, p. 110). They follow what is sometimes called "the rationalist´s strategy", which has been heavily influenced by military experience, where the strategy, in principle, consists of the following steps (Tidd et al. 2001, p. 70):

- Describe, understand, and analyze the environment
- Determine a course of action in the light of the analysis
- Carry out the decided course of action

This strategy, however, often does not work for complex and fast changing environments typical to innovation.

Doing things right the first time is possible only if most of the important facts are known and no vital disturbances occur during the work to finish a task. However, both of these necessary demands are seldom at hand, especially for innovation projects in their early stages. Without this very important knowledge, financers and board members will make the wrong demands to the entrepreneurs and the development teams. In turn, the entrepreneurs, living with the wrong expectations, will spend unnecessary time and resources trying to get proper decision material to be able to "do it right the first time". Their only possibility to find the reliable answers is to test different solutions and learn from the tests and to do new tests in an iterative way.

Therefore, early in the development of innovations, the situation is very different from the classic management and development principles that have been developed for them. Thus, the managers and team workers often have to rely on no, very little, and/or unreliable information in a dynamic/agile reality when deciding on an action or taking a solution in order to continue their work. There are cases where success comes by accident and sometimes the benefits arising from one lucky break are enough to cover several subsequent failures. In a process as uncertain and complex as innovation, luck also plays a part (Tidd et al. 2001, p. 18), as does timing as well.

Until a stable product and market situation has been reached—and when unplanned situations occur—it will feel like walking through a labyrinth (see Fig. 2.7). Every time the entrepreneurs come to a dead end, they will get an unpleasant, frustrating and stressful feeling, although going back to start over again normally means that knowledge from the first attempt can be reused to bring the process/product to a fast and efficient end. Therefore, the finding of a well-functioning concept for innovations is a gigantic learning process, characterized by "learning by doing", "learning by testing", and even "learning by failing". Every time a dead end situation is reached, it is leadership that is the critical factor for whether or not the project will come back on track.

The labyrinth metaphor is an example of "the incrementalist's strategy", characterized by the following procedure (Tidd et al. 2001, p. 73):

- Make deliberate steps (or changes) towards the stated objective
- Measure and evaluate the effects of the steps (changes)
- Adjust the object if necessary and decide on the next step (change)

Thus, in conditions of complexity and change—in other words, the conditions for managing innovation—there are no easily applicable recipes for successful

Fig. 2.7 An innovation project is like finding the way through a labyrinth. The way to get through it is unknown until the way through has been found

Fig. 2.8 In an innovation project, important puzzle pieces are often missing and they must be found with partly irrational methods that get information by chance

management practice (Tidd et al. 2001, p. 74). Instead one has to rely on what might be called an innovation philosophy.

Passing through a labyrinth by using more or less rational reasons for the next step must also be complemented with the finding of missing "puzzle pieces" (see Fig. 2.8). The finding of missing puzzle pieces is often shown to have been the result of coincidence and unplanned meetings with unknown people, for example, in airplanes, busses, or trains. A great help to being 'lucky' is being curious and open minded when searching for missing puzzle pieces.

By talking and mingling with known and unknown people and by reading, listening to and viewing different news media, useful information often pops up that can be of immediate use or later use in an innovation project. Random searching, for example on the Web (i.e., World Wide Web—www) and especially with Google can also result in similar, unintentional payback.

2.6 Innovation Projects

For the development of a new innovation, an innovation project needs to be set up. In that process, the first thing to do, ideally, should be to appoint a leader—the entrepreneur, the intrapreneur, or the renovateur—of the innovation project. She/he should then be responsible for the recruitment of the people or consultants needed for the project.

For the innovation project to be successful, the innovation project leader (hereafter called entrepreneur) often needs to arrange sub-projects, such as taking care of product development with supply chain development, sales development, and market development (see Fig. 2.9). For the administrative work to be successful the entrepreneur normally needs a project assistant or a project staff for larger projects (named Administration in the Figure). The over-all business demands that the entrepreneur has to keep in mind, normally need to be changed a number of

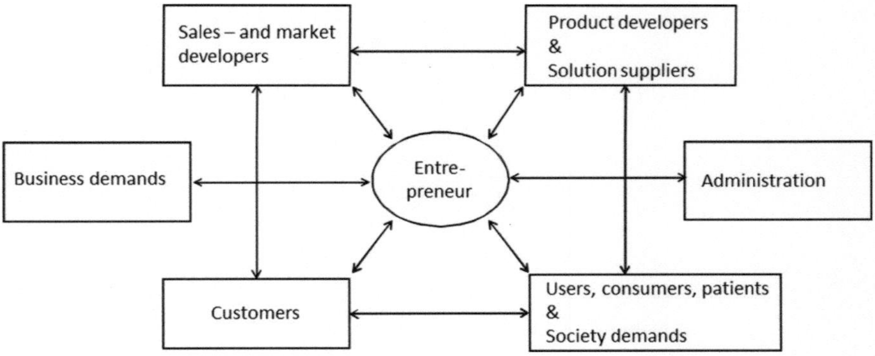

Fig. 2.9 Innovation development, orchestrated by the entrepreneur, takes place in a complex, adaptive social system

times, based on changing circumstances and what has been learned from the actual project work.

The salesmen and market developers must be closely related to the potential customers, while the product developers and supply chain developers must be closely related to the users and consumers of the products being developed. As a number of interrelations exist in an innovation project, this makes it an example of a complex, adaptive social system.

Supply chain activities transform natural resources, raw materials and components into a finished product that is transported to a place of delivery (see Fig. 2.10). By utilizing advanced information and communication technology (ICT), a highly efficient total supply chain process is often achieved. In sophisticated supply chain systems, used products may re-enter the supply chain at any point where their residual value is recyclable. As seen in the Figure, customers of the product can be citizens (individuals), businesses, and governments, as well as non-profit organizations. When Artificial intelligence (AI), intelligent robotics, and the Internet of Things (IoT) develop, we also get an increasing situation with non-human customers. The three customer categories in the Figure correspond to terms used in e-commerce.

For software products, the supply chain is normally simple as most of the distribution can be done over the Internet.

2.7 Creativity

Often creativity is needed to find suggestions for solutions to test when a problem occurs. Stimuli for creativity to take place can be the need to solve an existing problem (a *need*), a constructed problem (a *want*), or a vision (a *wish*). Other stimuli are, for example, to impress other people, to be unique, or simply to have fun doing things in new ways. Important input for creative/inventive work is information from

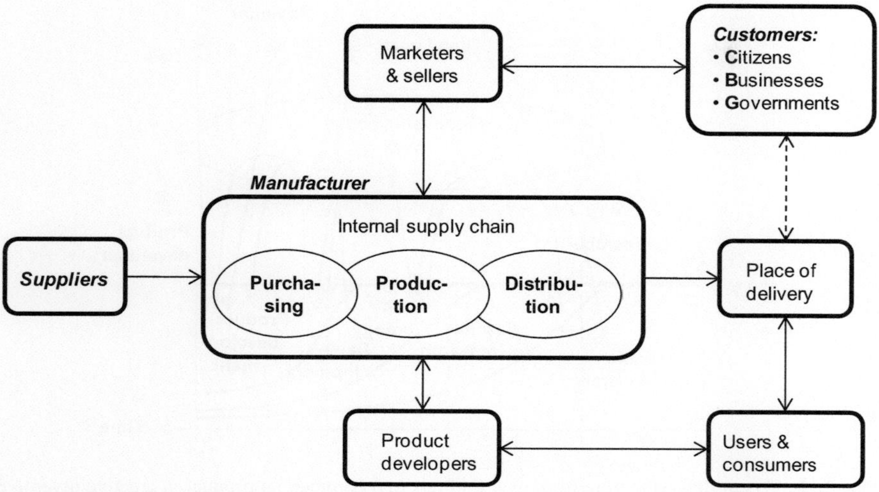

Fig. 2.10 An example of the supply chain for hardware products (inspired by Chen and Paulraj 2004)

observations, one's own use, and other information from solutions and findings in basic research, applied research, and technological development.

Generally seen, creativity is the capacity to perform creative acts and a creative act is an instantiation of the general process of creation. Creation represents a radical break with the past and the emergence of something that has had no prior existence. Creation also disturbs the balance and violates the status quo (Shkliarevsky 2017).

To be creative, the individual benefits if she/he has memorized a lot of different solutions and if she/he also has critically evaluated each of them, at least sorting them out as good and bad. By being curious and by being in in contact with a constant (massive) flow of information of new solutions and findings, the capacity for solving new problems will increase, both alone or with others. By sharing creative thinking in dialogues with other creative people, the solutions, in general, will be better than if a person tries to find creative solutions on her/his own. Especially structured creative meetings, such as brainstorming, which is treated later, can help to get creative ideas to work within an hour or so. Then, the time it takes to transform an idea into a commercial product will also, generally, be shorter if some creative people cooperate instead of if the idea owner tries to develop the idea isolated from others.

To cut down the time from the appearance of a new idea to the launch of a new product on the market, creative researchers, inventors, lead users and product developers can contribute a great deal if they interact with each other in what can be called a creative network (see Fig. 2.11). Eventually Information and Communication Technology (ICT) possibilities can be used to form virtual e-innovation centers. The "short-cut" can be done in steps, for example, from basic

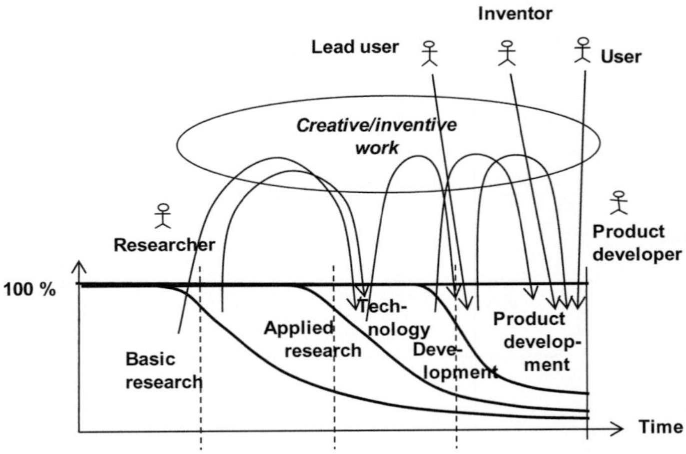

Fig. 2.11 To cut down the time from new findings to a commercial product, a creative/inventive network can help the idea owners to speed up and improve the development process

research to applied research to product development. Note that, in this respect, it is important that patent applications are filed before the researchers publish their scientific results in academic and other publications. Previously published results, or results presented in public can create difficulties in the acquisition of patents.

2.8 Product Values

The simple question 'what is a product?' is not so easy to answer. For example, is a nice stone that nature has shaped a product if someone has picked it up and tries to sell it? Is there a difference if the same person or company sells many of the same stones? Probably most of us will accept that a stone that has been treated in some way is a product if it is sold on the market. But, how much treatment of the stone can we accept for it to be regarded as a product? A "manmade" product, perhaps not intended to be sold, can also be called an *artifact*. Still another man made product that is only produced to be sold in a limited number of copies by an artist is called a piece of art. But, what if the artist gets an order to make a piece of art that is duplicated for a company to be given away as gifts? In that case, is the artifact a product or should it, perhaps, be called a service?

In general, 'product' in this book means hardware and/or software. Hardware and/or software can have different degrees of complicatedness and one classification can therefore be; simple, complicated, complex, and complex adaptive. Another classification can be low-tech, medium-tech, and high-tech products. Also products can be primary and secondary products, meaning that the main product is the primary product and the secondary products are parts and systems that are needed for the function of the primary product. Examples of secondary products are supports, petrol, manuals, educational material, packages, or service materials.

Complicated, complex, and complex adaptive products are, in general, high-tech products. A significant difference between complicated and complex products is that the complicated products are mostly passive, while complex products are reactive, such as is the situation for the Internet-of-things (IoT) or Machine-to-Machine (M2M) solutions. To be able to be reactive, the products must have sensors in order to get input to react on. They also must have built in programmed devices, such as micro computers or bi-metals to be able to change or make things on their own when different, pre-planned situations occur. Mechanical robots and maybe, although emotionally tricky, trained animals, such as dolphins and watchdogs belong to reactive products.

In earlier times, slaves were sold as products, which unfortunately is still a problem with human trafficking. Products that are complex adaptive have sensors as built in (artificial) intelligence so that they are able to adapt themselves to a new situation. One such product is the software that finds the pattern of a person's purchasing profile and suggests to the same person to buy another new, similar product.

Seen from a user's point of view, a product has a technical content and has, if one in any way engages oneself with it, also a mindful content. [Mindfulness is the psychological process of bringing one's attention to the internal and external experiences occurring in the present moment (Pagnini and Phillips 2015).] Under the two content terms, at least eight product values can be listed, and these can overlap each other:

Technical content

- *Functional values* are dependent on the technical solutions mostly hidden inside the product. The function can be as simple as just filling in the space (e.g., a gas in a balloon or concrete in walls). It can also be advanced, with all degrees between simple and advanced, for example, an engine in a car has simple as well as advanced parts and systems
- *Interaction values* determine how well a user can interact with a product
- *Sustainability values* are long lasting, environmentally responsible values for the users, society, and the providers (the business)

Mindful content

- *Perceptory/sensorial values* are based on what we experience with our five classic senses (see/hear/taste/touch/smell) from outside and/or in contact with a product. The product semantics is an important part of these values. Product semantics is an attempt to convey what a product is or does through its form.
- *Emotional values* are the passion/feelings we have for a product. The product semantics can also influence these values.
- *Image values* are based on the image we get of the product and what we think of it, for example, when closing our eyes. Brand names, patents, the image given on web pages, stories and the expressed experiences of the product by other

users will influence and develop the image we have of a product. The product semantics can influence these values.
- *Stimulation values* stimulate wellbeing and make life easier to live. They can be training or problem solving activities. such as a game, an e-commerce possibility, or simply robots and products with built in AI (Artificial Intelligence) and IoT (Internet of Things) applications that have to be programmed, monitored and interacted with in order to do the needed, wanted or wished job.
- *Existential values* are values that are experienced in extreme environments, such as intensive care units and remand prisons (Torkilsby 2014). The wellbeing in such environments is dependent on their design.

(There have been Five core interconnected dimensions of wellbeing mentioned (Rath, T. and Harter, J., 2010):

- Career Wellbeing: How you occupy your time
- Social Wellbeing: Strong relationships and love
- Financial Wellbeing: Managing your economic life to reduce stress and increase security
- Physical Wellbeing: good health and enough energy
- Community Wellbeing: Sense of engagement and involvement where you live)

In the classical industrial era of product development and engineering design, almost all efforts were made to take care of the functional values. For modern, new product development (NPD) the other soft values must also be taken care of if the new product is to become a successful innovation.

From the beginning of the 21st century, the concept of mass customization (e.g., Tseng and Jiao 2001) has spread, which means that the market-need view has been extended to include satisfying single users and customers with customized products based on modularizing principles. One reason for this is that sellers want to customize their product offerings in order to increase the value offered to individual buyers and to gain a competitive advantage over the their competitors (Stump et al. 2003).

With the development of the World Wide Web (www), a step further was taken to let the users design their products, for which there are two fundamental approaches (Randall et al. 2007): needs-based systems and parameter-based systems. With *needs-based systems*, users specify the relative importance of their need, and an optimization algorithm recommends the combination of design parameters that is likely to maximize the user utility. With *parameter-based systems*, users directly specify the values of the design parameters of the product.

Based on political initiatives to satisfy public health reasons and democratic needs, there is also a growing market for universal products, so that everybody can have equal possibilities to use the products and public spaces. The need for products and public environments with a high degree of universality is therefore based on a political want or wish (e.g., EU 2007). Figure 2.12 shows some characteristics of the three development situations with approximate years of initiation.

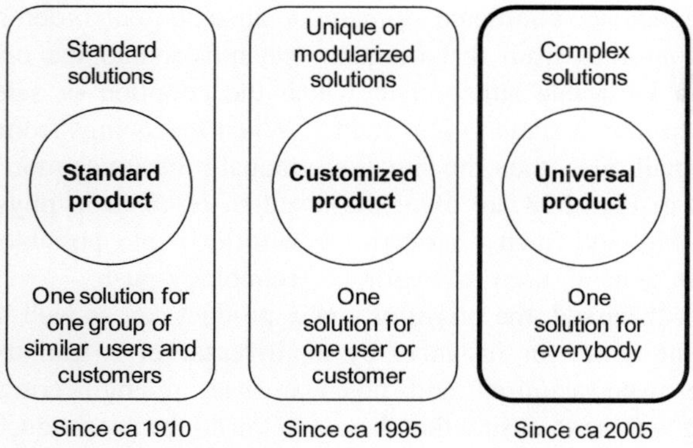

Fig. 2.12 Three product types (Ottosson 2013)

2.9 Product Life Cycle

Figure 2.13 shows the situation from the birth of a new product and business idea from the new knowledge situation in Fig. 2.1 to when the product is removed from production and sales for an individual producer. Until the sales and marketing people have managed to build up a critical mass of customers, everybody involved in the innovation project will initially experience a positive chaotic and often euphoric situation. With the initial experience of the market, the chaotic situation is gradually transformed into a complex situation, which means that one gets into

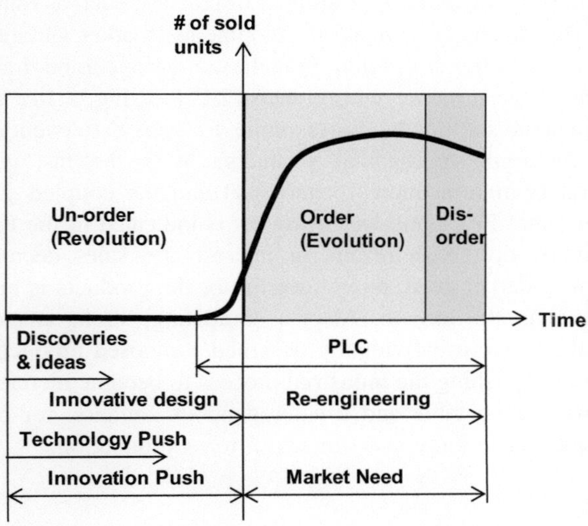

Fig. 2.13 The process from the birth of an innovation until it is removed from sales (PLC stands for the commercial product life cycle)

what can be called an unordered domain or an emergent order domain. If the development continues from that domain, one moves into the ordered domain, which means a knowable situation in which the adoption or sales/income that swings month to month starts to get smaller. When the swings from one month to the other are small ones, then the situation gradually becomes more stable.

Discoveries in Fig. 2.13 are often the result of research in physics, medicine, chemistry and biology. Such discoveries transformed into possible technological solutions are, in general, seen as creating a technology push.

As was also discussed, the origin of a new product can as well be the result of the development done, for instance, by an inventor or a lead user. When the products have been acquired and taken in use, re-engineering takes place. Re-engineering is done to ensure that the market need for the products is cultivated, as long as the products are not being regarded as old-fashioned by the market. In general, re-engineering is easy to plan and the returns on investments can be calculated, which is often opposite to the situation when innovative development is done.

When the growing interest and sales of the product starts to slow down, sooner or later one gets into the disordered domain characterized by different opinions about what to do to improve the situation. When the cash flow starts to be negative, the pressure increases on the business and an increased negative chaotic situation often becomes a reality. This negative, chaotic situation will also arise every time sales drops drastically over time. For a mature company to avoid the negative chaos becoming too deep, it is important to have started new innovation processes early. One or more of these can successively take over when the standard product is not possible to sell anymore, for example, when the end of the commercial product life cycle (PLC) is approaching.

The PLC varies significantly for a number of reasons, such as financial strength, willingness by the financers to make developments, market situation, or type of product. However, looking at different branches, a curve can be drawn comparing R&D in percent of the turnover with branch PLC (see Fig. 2.14). As shown, the PLC varies from about six months (for example, for games) to twenty or more years (for example, for some mechanical products). With the fast growth of ICT (Information and Communication Technology) and the coupled globalization, a trend is emerging that PLCs are decreasing, as is indicated in the Figure.

When the innovation push results in interest and sales decreasing and the establishment of a market need, re-engineering of the products is needed to avoid losing momentum in the growth. Also, a competing, strong market trend is to produce increasingly more individually designed/composed products. In turn, this results in the market pushing the industrial process to become increasingly flexible, causing an unwanted unstable and a heterogeneous situation for companies that have, for example, made large investments in mass production.

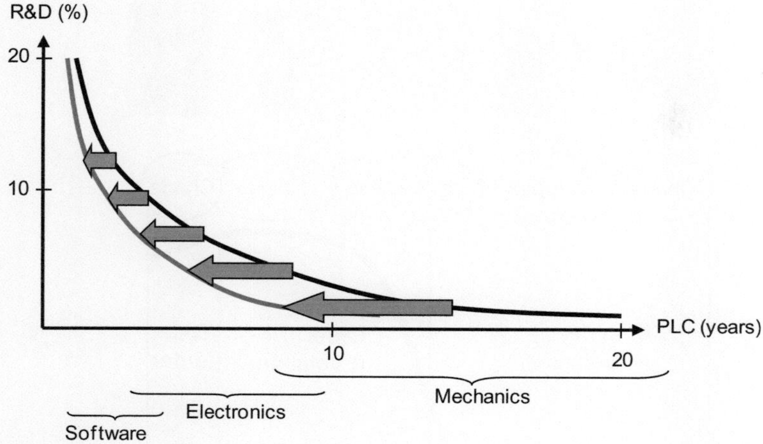

Fig. 2.14 PLCs vary between branches. There is a trend that PLCs are decreasing as is indicated in the Figure (Ottosson 2004-C)

2.10 Three Metaphors

The un-ordered part in Fig. 2.13 often starts with a euphoric positive-creative chaotic situation, which sometimes is called the Fuzzy Front End (FFE) of an innovation project. When one or more promising ideas/solutions have appeared, the first rough product concepts can be set up. For the materialization of the product concepts into new products, the project leader and her/his team are then faced with a complex situation, for which creativity and improvisation is needed as well as the need to test ideas, to immediately make changes and to make new tests. Doing so, they gradually build up knowledge about what works in reality, both from a technical and a use or a user perspective.

When a customer first acquires the new product and it has been taken in use—when it has become an innovation—a knowable situation is at hand. The more customers that acquire the product and the more users that use the product, the more known the situation will be for and around the product. If the further development of the product and services around it is not good enough, the interest and sales will go down, causing negative chaos to appear. Thus, the five different stages from the birth of the business/product idea to when it is not sold any more can be called: (positive) chaos, complex, knowable, known, and (negative) chaos. The first three stages have similarities with playing dice, poker and chess (see Fig. 2.15).

Note that the "S curves" in real innovation projects are not so smooth as shown and that in subprojects, a reached knowable situation can quickly be turned into a new, complex and even chaotic situation. However, seen from above, when different subprojects are summarized, an "S shaped" curve can sometimes be constructed. It also has a pedagogic advantage to use S-curves to explain processes, which is maybe the most important reason why they are often used.

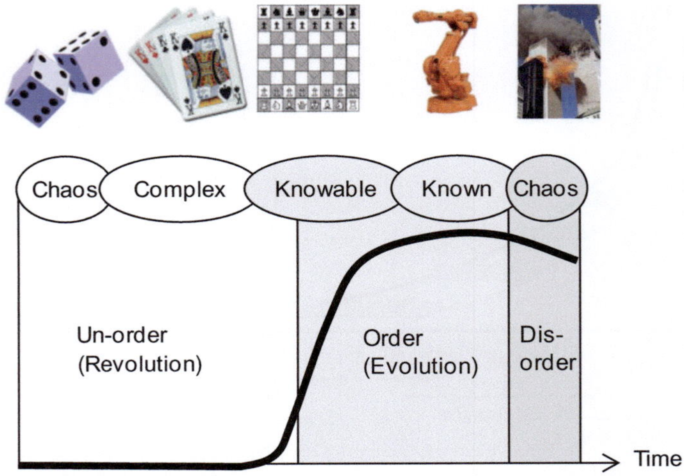

Fig. 2.15 Different realities with metaphors for the early stages of the development

Dealing with chaos, complexity and knowable situations is, in many ways, like playing dice, poker, and chess. Table 2.1 shows briefly how this metaphor can be applied when trying to inform about the different situations in the development of innovation projects.

The different situations for an innovation project and standard business processes can be broadly summarized, as is shown in Table 2.2. (C&E = Cause and Effect. In the transformation, many new realities will be experienced by the people following the development).

Table 2.1 Differences between playing dice, poker and chess (Järrehult 2006)

Chess (need) 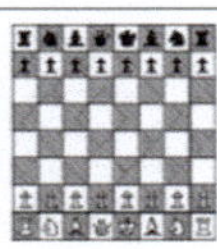	Poker (want) 	Dice (wish) 
You plan way ahead	You plan short term but think long term	No planning possible
You have most of the information needed	You pay for new information	You get no new information
You know what you have got and what competition has	You discover along the road what you and others have and get	No cause effect relationship
Slow games	Fast games	Very fast games
Low risk taking ability	Larger risk taking ability	Very big risk taking
Often duce	Several winners possible	Stochastic outcome

Table 2.2 A successful innovation project transforms into a business process (based on Kurtz and Snowden (2003))

	Innovation project		Standard buisness process		
	Un-order (emergent order)		Order		Dis-order
Reality	Positive chaos	Complex	Complicated	Known	Negative chaos
Cause and effect (C&E)	No C&E relationship perceivable	C&E are only coherent in retro spect	C&E separated overtime and space	C&E relations repeatable, perceivable and predictable	C&E understand-able and choking
Decision input	Ad-hoc actions	Pattern seeking	Analytical/ reductionist investigations	Follow ups and best practise	Ad hoc actions
Planning	Day-to-day	Short term	Short and long term	Long term	Short and long term
Actions	Act–sense-respond	Probe–sense-respond	Sense–analyze–respond	Sense–categorize-respond	Act-sense-repond
Management	No	Entrepre-neurial	Trans-formative	Administrative	Change
Organizatio	Informal	Planetary	Matrix	Line	Planetary

2.11 New Products

The newness of a product can be seen from two perspectives, that of the producer/supplier and that of the market (see Fig. 2.16). As newness can be difficult to measure, there is a rule of thumb that says that the new technical content should at least be 60% compared to the existing products of the company (by "new content" we mean newly developed or re-engineered details and/or systems). However, the perceptory values must also be changed if the product is to be marketed as a new product, in order to give the customers/users/consumers the feeling of newness. If, in addition, the brand of the product is also changed, the market can maybe be convinced that the product is really new.

If, for example, a new refrigerator has the same technical solutions as another refrigerator but is painted in a new color, it does not fulfill the demands to be a new product. Also, if the color of the new refrigerator is not changed and if the same disposition of the cabinet is used as for existing refrigerators but the whole cooling system is changed, it is not a completely new product.

If the technical content is completely new, a patent can eventually be secured for the solution, giving the company a monopoly situation on the market for some time (<20 years). As stated earlier, patents belong to what is called *Intellectual Properties (IP)*. Products that can be patented are often called inventions and the people who have been filed as creators are called *inventors* (not to be confused with inventor buyers). Inventors can be professional inventors who live on the royalty

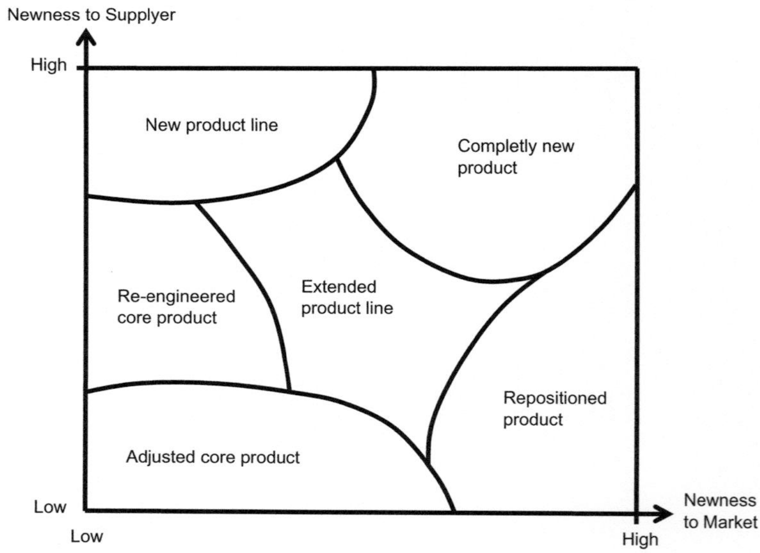

Fig. 2.16 A mature larger supplier will normally have products of different levels of newness. These products will be considered to have different newness by the market (from Wikipedia August 2007)

Table 2.3 Levels of novelty in investigated patents given during the 1970s and 1980s (Altshuller 1980)

Level one	Routine design problems solved by methods well known within the specialty. No invention needed. About 32% of the solutions fell into this category
Level two	Minor improvements to an existing system, by methods known within the industry. Usually with some compromise. About 45% of the solutions fell into this category
Level three	Fundamental improvement to an existing system, by methods known outside the industry. Contradictions resolved. About 18% of the solutions fell into this category
Level four	A new generation that uses a new principle to perform the primary functions of the system. Solution found more in science than in technology. About 4% of the solutions fell into this category
Level five	A rare scientific discovery or pioneering invention of essentially a new system. About 1% of the solutions fell into this category

income from their inventions. Royalty means some dividend of the sales price of the product. They can be lead users (inventive frequent users) and what can be called occasional inventors or amateur inventors.

Based on extensive investigations and classifications of patents and other solutions, Altshuller (1980) showed that the novelty can basically be classified into five levels (see Table 2.3).

Of great importance for all involved in a product development project is to always keep in mind that it is advantageous to try to find solutions for which the users can use already engraved/acquired behaviors. Not only will the users find a greater pleasure if that is done, but it has a definitive positive influence on the sales and profit figures for the company. Figure 2.17 shows this in a schematic way.

2.12 Market Adoption

In economics, market saturation is a term used to describe a situation in which a product has become diffused (distributed) within a market. The actual level of saturation depends on many factors, such as customer purchasing power, competition, prices, and technology. When the saturation rate for a solution is close to 100%', the market is said to be saturated (see Fig. 2.18) as further growth of sales will occur only as a result of population growth. For example, in advanced economies, an extremely high percentage of households own refrigerators (more than 97% of the households), which means a saturation rate more than 97% in this case.

To the left in Fig. 2.18, typical adopting categories are mentioned and the order in which they appear. Of the different categories, *Pioneers* and *Early adopters* are visionaries. They want revolutionary change—something that sets them apart qualitatively from later buyer categories (Gladwell 2000, p. 198).

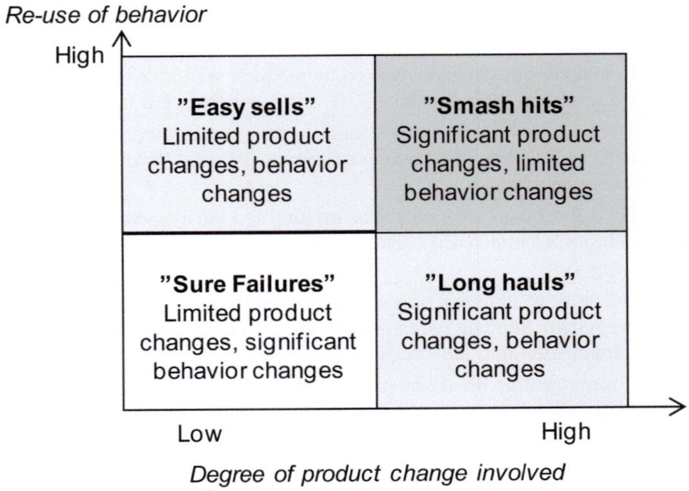

Fig. 2.17 The use of engraved behaviors is positive for the success of a new product (Gourville 2006)

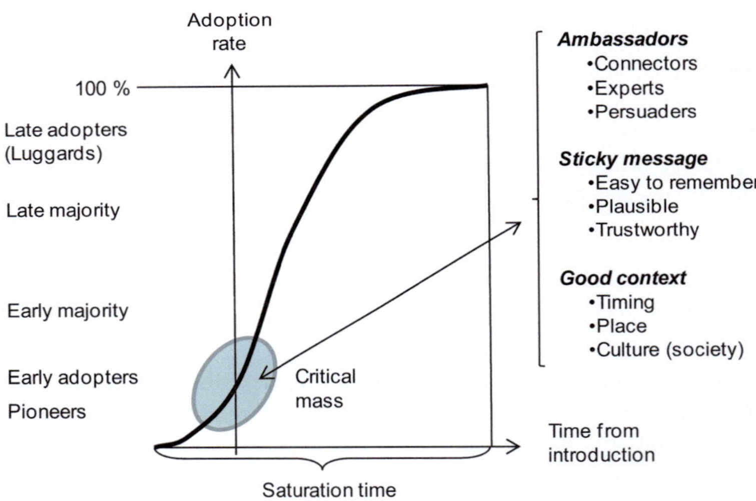

Fig. 2.18 The principal adoption curve for an innovation

When a critical mass has been reached for an innovation, many individuals—ambassadors—will help to amplify the further rate of adoption. For rapid growth, sticky messages and fertile contexts are needed. What characterizes sticky messages is that they are memorable, plausible and trustworthy. A fertile (or fruitful) context means good cultural conditions and political circumstances, good timing, and more.

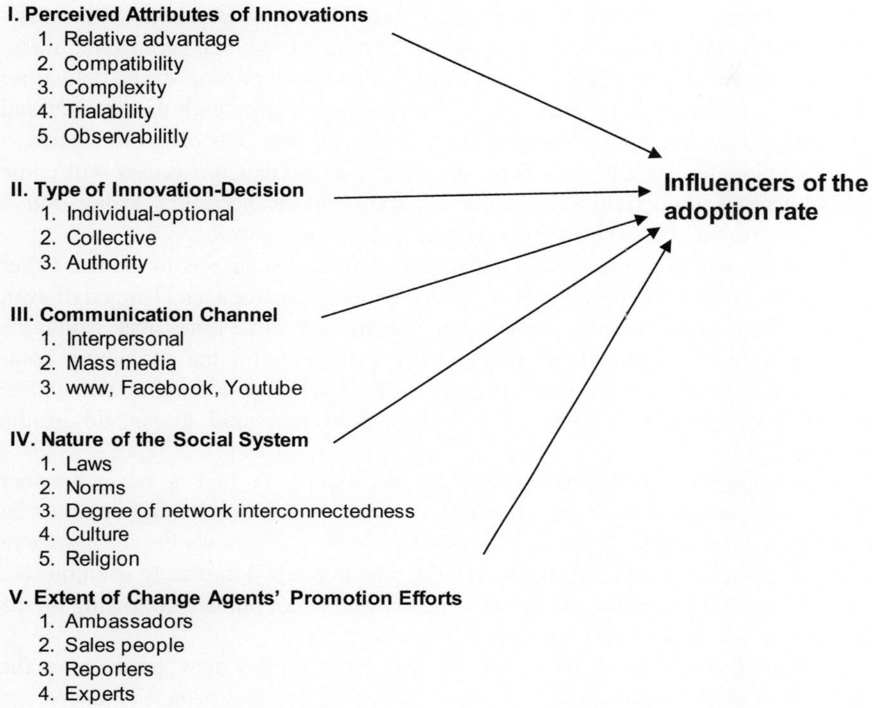

Fig. 2.19 Examples of variables determining the rate of adoption of innovation (based on Rogers 2003, p. 222)

The ambassadors in Fig. 2.18 are both people who have bought and/or used or consumed the products and connectors, experts, and persuaders (Gladwell 2000, p. 33). The ambassadors are, in general, unpaid but their contribution to the sales growth is often ego supportive, for example, to be proud of someone asking for their knowledge, or to be mentioned. A special type of ambassador is the bloggers who are also called *influencers*. Such influencers are people who, through a well-visited blog, Instagram account or Youtube channel, have a big influence on social media.

The *connectors* know lots of people and are good at networking. The experts are "information distributors" who are knowledgeable people. The persuaders persuade us when we are unconvinced. Ambassadors also use social media to spread their messages, which often makes them spread quickly like a virus.

On the same theme, Rogers (2003) gave five variables; Perceived Attributes of Innovations, Type of Innovation-Decision, Communication Channel, Nature of the Social System, and Extent of Change Agents' Promotion Efforts, They determine the rate of adoption of innovations, meaning number of individuals that adopt a new idea in a specified period. Figure 2.19 specifies each variable in more detail.

Until a critical mass has been established, innovation push is needed. This means active market activities, such as searching for potential customers, contacting them, informing them, convincing them, negotiating with them, closing deals, delivering the product, teaching them, and making sure they are happy with the product and service they have received. From the time the critical mass has been established, a viral (or epidemic) growth can occur, meaning that spontaneous orders will come without active sales activities. Now that a market pull has become a reality, active sales activities can be stopped without orders stopping as well.

When take-off takes place for a radical innovation, it soon gets followers. Other entrepreneurs will think they can also be successful with somewhat changed or even better solutions than the initial innovation. When, after some time, many solutions are competing, the profitability situation will get harder for the competing companies. Around this time, some companies will go bankrupt and some will start to acquire other companies to get a bigger market share and to get rid of the competition.

A consequence of this turbulence in the market is that a normalization/ standardization will take place. After some time, only few dominating commercial solutions will be available for the customers. Figure 2.20 shows the development from the first revolutionary solution until the time that few dominating solutions are available. Recent examples are iPhones and android smart phones, following up the smartphone revolution stated by Steve Jobs and Apple.

In general, sales are initially slow when introducing a new product on the market. Not until the critical mass of users and customers has been established can the ambassadors generate new customers in an efficient way. This applies, above all, to groundbreaking new products since customers rarely end up to be the first to

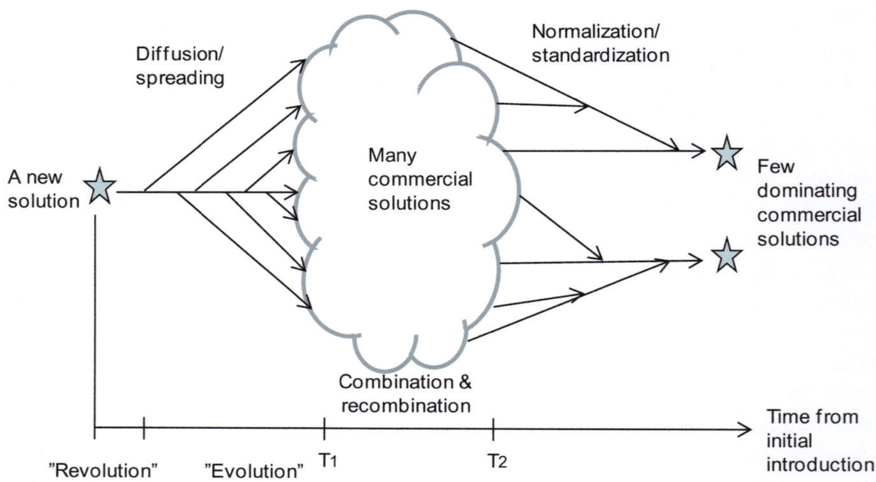

Fig. 2.20 From a single revolutionary solution, many solutions will develop, ending with a few dominating solutions

buy a new, unknown product with no reputation. Therefore, in order to start sales, one needs to be highly active in networking and in the making of market contacts. Often this will feel like an assault on an unconquerable mountain. Once one has learned how the product is to be sold and sales achieve a lift-off, a relatively fast rise in sales is possible. A metaphor of this is shaking a ketchup bottle to at last get the ketchup to come out in a massive flow.

When one has a feeling of euphoria at having succeeded, and growth is proceeding positively, one can suddenly run into the doldrums, with sales plummeting. Then it is natural to look for logical explanations for why sales have dropped off. Is the price of the product too high? Is the design wrong? Is the marketing wrong? Is the marketing material or the web page bad? Has the company the wrong sales personnel? Are the sales personnel lazy? What is wrong? Questions abound, and panic can grow as various countermoves fail to produce noticeable results. In a small company, the drop in sales means that liquidity problems arise as the fixed costs begin to exceed sales revenue.

There can be many reasons for such a sudden sales drop. One reason can be, for example, that the early pioneers note the functional values of the new product above all and maybe not so much the sensorial/perceptory values. When it is time for the later pioneers—the early adopters—to acquire the product, there is also a demand that the product has good sensorial/perceptory values. In addition, during the initial stage of marketing, direct sales often take place. In the subsequent stage, more people can be involved in the buying process, thus slowing it down. The slowest buying process seems to be when authorities are involved.

2.13 Customers and Users

In general, the classic theories do not distinguish between users, consumers, and customers. However, it is important to point out that a user uses a product over and over again, while a consumer consumes a product once (e.g., a hamburger) or a number of times during a limited time (e.g., a piece of music downloaded to the smart phone from Spotify). Mostly, the customers who pay or authorize payment for a product or service do not themselves use or consume the products. This has a vital implication for any innovation activity as the product developers must concentrate on satisfying the needs, wants or wishes of the users' or consumers' while the marketing and sales people must concentrate on the customers.

Therefore, as a general recommendation, the product developers should be in close contact with the users and consumers. However, letting the product developers be in close contact with the customers often reduces the salespeople's (or seller's) possibilities to make optimal deals when the customers are not also the users or consumers of the products.

In general, users can be divided into different target groups depending on the aim and the situation. With what is called Universal Design (see below) or Design for all (Dfa), the idea is to make products for all types of human users including ordinary users, users with special needs, and disabled users.

What characterizes the three user groups are (Björk 2003):

- *Ordinary users* are able bodied, capable individuals who are in a position to freely choose what kind of product or service they want or need. If necessary, ordinary users can to some degree adapt themselves both physically and mentally to poor technical solutions.
- *Users with special needs* are individuals who have no recognized physical or mental impairment but who need some special modification of standard products in order to use them properly. For example, elderly people with minor disabilities, such as reduced strength, or impaired hearing belong to this group.
- *Disabled users* require assistive devices due to severe mobility problems and reduced body functions. The products must have a high degree of both technical and practical functionality; otherwise the products will not work as intended. Disabled users do not have the freedom of choice enjoyed by ordinary users, as they cannot compensate for practical disadvantages in the same way as ordinary users. This group of users is heterogeneous and consists of individuals who have higher individual demands than ordinary users.

For computer—and Internet-based products it is possible to make products that will work for almost all of these three user categories, but it often becomes difficult to make mechanical products that will work for all three types of human users (Björk et al. 2007).

The Committee of Ministers of the European Union decided on the 12th of December 2007 (see Resolution ResAP(2007)3, to recommend to the Governments of the EU states to accept "Universal Design" as a philosophy and strategy supporting implementation of full citizenship and independent living for all people, including people with disabilities. According to the resolution, "Universal Design is a strategy which aims to make the design and composition of different environments, products, communication, information technology and services accessible and understandable to, as well as usable by, everyone, to the greatest extent in the most independent and natural manner possible, preferably without the need for adaptation or specialized solutions."

Furthermore, the Committee of Ministers of EU claims that "The aim of Universal Design is to make the built environment, communication, products and services accessible and usable to the greatest extent possible. It promotes a shift towards user-centered design by following a holistic approach and aiming to accommodate the needs of people with disabilities, regardless of any changes they might experience in the course of their lives. Consequently, Universal Design is a concept that extends beyond the issues of mere accessibility of buildings for people with disabilities and should become an integrated part of policies and planning in all aspects of society."

Seven principles for Universal Design have been suggested (Story et al. 2001) to make sure that high usability is achieved for hardware products. However, for software products, Principle 7 and part of Principle 6 below are not applicable:

1. Equitable use
2. Flexibility in use
3. Simple and intuitive
4. Perceptible information
5. Tolerance for error
6. Low physical effort
7. Size and space for approach and use

We can also add another two relevant principles when users pay for the products they will use (Björk & Ottosson 2016). In many countries, the customers are the public sector representatives, such as municipalities and governmental social security bodies.

8. Comfort in use—from comfort to discomfort to pain
9. Joy/pleasure in use

The disabled user group is not small and unimportant for product developers and it has been estimated that between 10 and 15% of the European population belongs to this user group. However, disability statistics tend to reflect users who are registered as disabled. Many disabled people do not consider themselves to be disabled or prefer not to register as such. The consequence of this is that disability statistics only approximate the true numbers and almost certainly underestimate them.

The disabled user number increases with the age of the citizens, although all of us now and then can be disabled (see Fig. 2.21). A and B in the Figure means that someone is injured or has been sick for some time and is having a recovery time to get back to a normal situation again. With special treatment, such surgery or training, a person can reach better situation, as shown by arrows C and D.

When extensive adaptation work is needed to make a product function well for people with special needs or for a disabled user, *assistive technology* is needed. For such adaption work to be acceptable or good requires a combination of technical and medical knowledge by the engineers.

Users can also, from time to time, be either primary, secondary, or co-users defined as:

- *Primary users* are individuals who use the products/services themselves.
- *Secondary users* are users who only occasionally use the product.
- *Co-users* are people who are involved in someone's use of a product without using it themselves (e.g., a driving instructor).

Figure 2.22 shows a summary of the different categories of users

The categorization of users as primary users, secondary users, and co-users is relevant to the forming of different product requirements for the various user

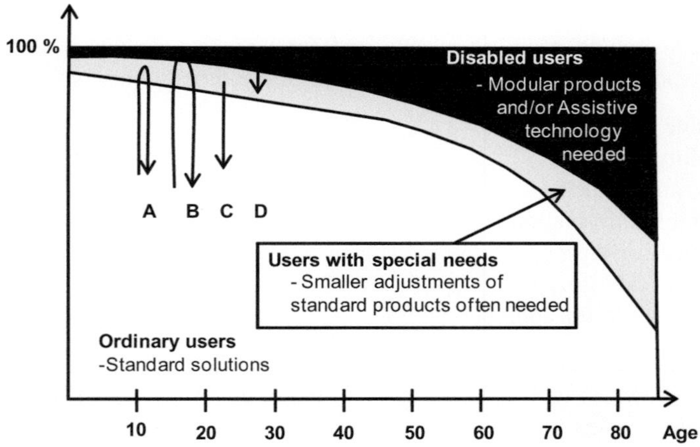

Fig. 2.21 The population can be divided in three broad user groups for which different standard product solutions have to be adjusted or adapted (based on U.S. Bureau of Census 1997)

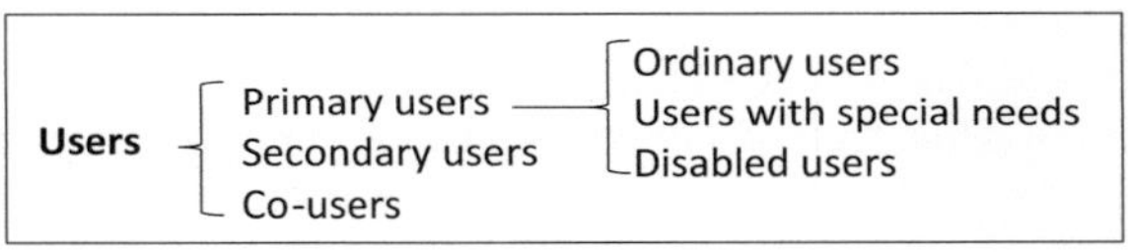

Fig. 2.22 Different user categories

groups. Especially in the assistive technology field, it is often important to be aware of the co-user demands as they sometimes handle the products even more frequently than the primary users do.

One example can explain the importance of the differentiation in primary, secondary, and co-users (see Fig. 2.23). A parent who frequently secures a child in a child-seat in the family car is an example of a co-user, while the child is the primary user. The parent focuses on the safety and comfort of the product as well as on a reasonable price and an attractive design. The taxi driver who takes the child to school normally is a secondary user. The taxi driver requires a product that can be installed quickly, that withstands dirt, that is light weight, easy to clean, and that takes little space when stored. Thus, parents and taxi drivers have different demands, which are important to consider in order to achieve optimal product usability.

2.14 SWOT Analyses

There are many solutions to compete with on the local and global markets (*glocal = global and local*). To try to find out if an innovation project will have good opportunities to be successful, SWOT analyses can be helpful. To lower the risk of

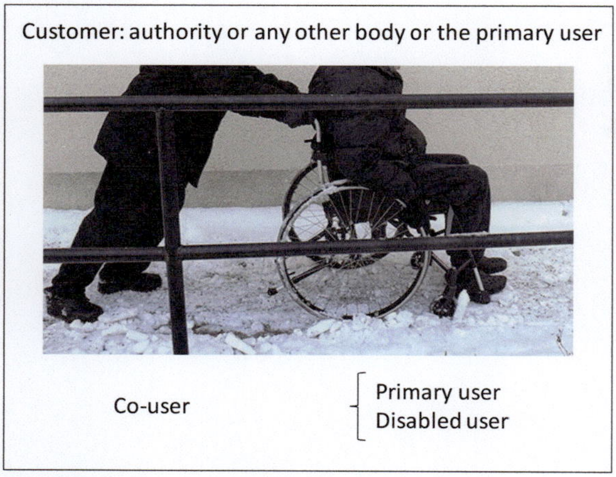

Fig. 2.23 An example of primary, secondary, and co-users

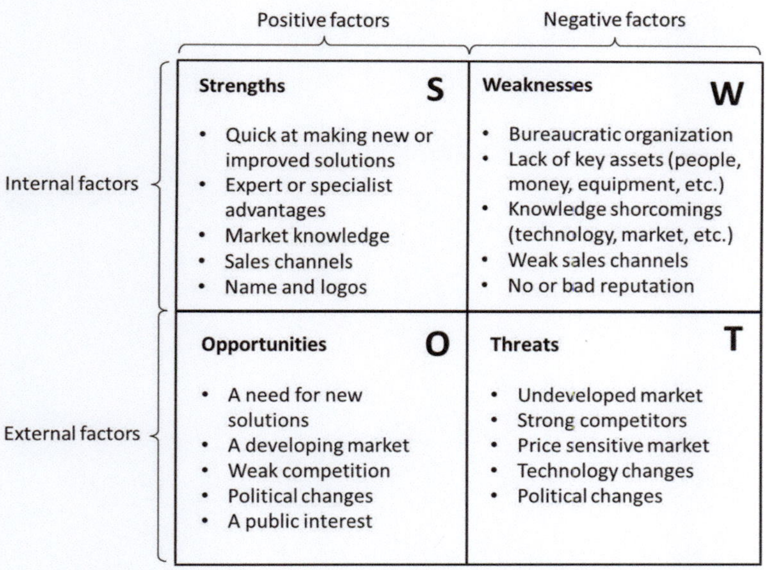

Fig. 2.24 Examples of factors that will influence the possibilities for success with an innovation project

failure, they should be done before the start of the project and then be updated when new information is at hand that might change the opportunities for success. SWOT stands for Strengths, Weaknesses, Opportunities, and Threats. Figure 2.24 gives some examples of SWOT parameters.

Chapter 3
The Entrepreneur

3.1 Inventors, Innovators and Entrepreneurs

As mentioned earlier, the project leaders of innovation projects are often called entrepreneurs, which is a short expression for innovation entrepreneurs (there are many types of entrepreneurs but the situation will generally tell what kind of entrepreneurs are meant). The term has a French background and we also use the terms intrapreneur and renovateur when discussing innovation leadership.

The important role of entrepreneurs in society is to create new ways of working and new businesses and activities with their innovation projects. There are similarities with the regeneration of old companies and activity units in the society, which is called change management performed by renovateurs.

If an entrepreneur does not manage to build an economical stable business, she/he is not regarded as a good entrepreneur. Unfortunately, this is a rather common situation especially if entrepreneurs have managed to get so called seed money from other people, such as business angels and venture capitalists. Thus, they are in a situation characterized by the fact that they do not need to push sales of the products, which is why sales will have difficulties to take off.

When inventors who make new solutions also develop and sell their new products, they can be called "innovators". What distinguishes an innovator from an entrepreneur is that the innovator must have some technical skills while the entrepreneur may not need them. Both types need to have a good enough practical knowledge of business administration and the ability to balance incomes and costs successfully so as not to put the business in financially critical situations. They also must understand basic business skills, such as sensible accounting practices, legal issues in the environment, in which the business operates, and the daily mechanics of financial and people management. When an organization develops an innovation, the term *business innovator* is sometimes used for that organization.

Most of the factors that the leader of an innovation project have to deal with are 'soft' and difficult to measure and handle. Therefore, having a good "business"

S. Ottosson, *Developing and Managing Innovation in a Fast Changing and Complex World*, https://doi.org/10.1007/978-3-319-94045-8_3

plan, a well thought business strategy and the "right" products will not guarantee success if the "wrong" entrepreneur has been chosen. An example of an opposite situation was at hand when the innovator of the Ice Hotel (www.icehotel.com) in North Sweden presented his 'crazy' business idea. He had only ice to work with to re-build his hotel once a year. The bankers did not want to lend him money to start the business as he could not give any securities in a hotel that melts down. Today, the business is a multi-million business!

The force that drives an innovator and an entrepreneur to start a new venture can have three principally different sources/reasons that can be denoted as opportunity, own talent and spiritual (Zohar 1997). Thus, the would-be innovator or entrepreneur looks at the market and notices an opening for some service or product and says; "Someone needs to provide this. I will". Further, the would-be entrepreneur or innovator looks inward at her/his personal resources and skills and says; "I can provide this". Finally, the future entrepreneur does not begin by thinking about a business or a career but about a feeling of inner necessity; "This has to exist. This has to happen. I have to do it".

3.2 Entrepreneurs in Different Roles

Depending upon the situation in question, the term "entrepreneur" can be divided into a number of sub-groups that are sometimes used independently in the debate, which is not correct. Earlier the three terms entrepreneur, intrapreneur, and renovateur have been mentioned briefly.

Principally, an (independent) entrepreneur starts her/his own business and often becomes the owner of it, at least during the start-up. Intrapreneurs start new business ventures for their employers. A renovateur is a person who saves businesses in trouble and who starts new ventures based on the best pieces of an old one (Bjerke 2013 p. 90).

The intrapreneur often has some limited topics to deal with as she/he must use the existing sales channels in the (larger) company, the internal accounting department, the internal human resource department, and more. Thus, while the entrepreneur has the full range of activities to handle, the intrapreneur has fewer activities to deal with.

Because of the urgent situation, the renovator must have the full power to act as does the independent entrepreneur.

Entrepreneurs have received a lot of attention in the literature. Intrapreneurs have also been of interest in the literature, while not so much has been written about renovateurs. Therefore Appendix B describes examples of the renovation of a business.

3.3 Managers and Entrepreneurs

The difference between an entrepreneur of any type, and a leader in general—a manager—is that the entrepreneurs build up a unique business while a manager looks for an existing business that is doing well. It is important in this work that the entrepreneur can both create confidence and build up a personal network, which in turn requires that she/he be gifted at networking and selling visions, ideas and products.

The pure entrepreneur's "compass" or "guiding rule" differs from that of the pure manager in many important aspects, as can be seen in Table 3.1. Note that the same lines of each column should be read against each other, so that the entrepreneur, for example, prioritizes acting/testing/trading and not much on planning, which is the opposite of that which the manager generally does.

The entrepreneur's and the manager's respective focusing on concrete action, planning and vision can schematically be expressed as shown in Fig. 3.1.

The entrepreneur has vision as her/his guiding star, set well ahead in time, and uses the vision for tangible acting/testing that provides immediate feedback. For the entrepreneur, planning, and waiting to receive feedback from the plans will be static, bureaucratic and non-creative. The manager, on the other hand, wants to see good order, and 'hates' chaos situations and disorder, which are very important ingredients in a creative and dynamic business and also give priority to planning before concrete acting.

Note that in this context, the focus on education at university colleges and universities has, in principle, traditionally adopted the profile of the manager, which stimulates the left cerebral hemisphere—the analytical cerebral hemisphere—of the human brain. Only in artistic educational disciplines, such as architecture, industry design, and artistry, acting in the right cerebral hemisphere is traditionally stimulated. The entrepreneur, however, must be able to move between creativity and analyze chaos and order, thus using both halves of the brain.

Table 3.1 This synthesis shows the entrepreneur's and the manager's guiding rules (Sjöstrand 1992)

Business logics of the entrepreneur	Business logics of the manager
Action—frequent testing	Planning
Vision	Information
Social cap ital	Financial capital
Cash-flow control	Return on investment
Network resources	Own resources
Confidence	Contract
Organization	Structure
Engagement	Judging

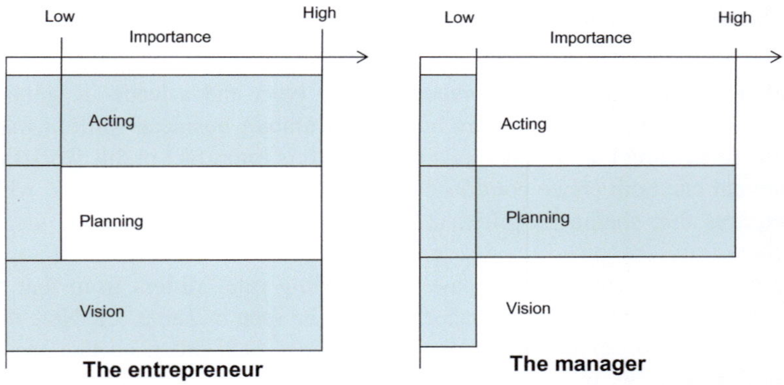

Fig. 3.1 The pure entrepreneur's and the pure manager's respective focus (Johannisson 1992)

Therefore, entrepreneurial leadership differs radically from the conventional leadership that dominates in leadership literature. Figure 3.2 shows a principle difference between the two.

Entrepreneurial leadership is primarily characterized by great flexibility and willingness to change, while the opposite is more applicable to the leadership of mature companies. To experiment and to learn from one's mistakes is a central part of entrepreneurial leadership, such as to quickly redefine and reorient the business as required by the development of events. Speed and timing are often of decisive importance to success or failure.

While managers are taught to analyze the market and choose the target segments with the highest potential return, entrepreneurs tend to find ways to reach the market with minimum expenditure of resources, such as time, effort and money (Sarasvathy 2001). Since entrepreneurs often tend to start the process without assuming the existence of a pre-determined market for their ideas, detailed competitive analyses do not seem to make any sense at the startup phase for them. Thus, they do not wait to "discover" the perfect opportunity. They make that opportunity effectually along with other effectuators, who self-select into the process of creating new opportunities, ventures, and markets (Sarasvathy 2006).

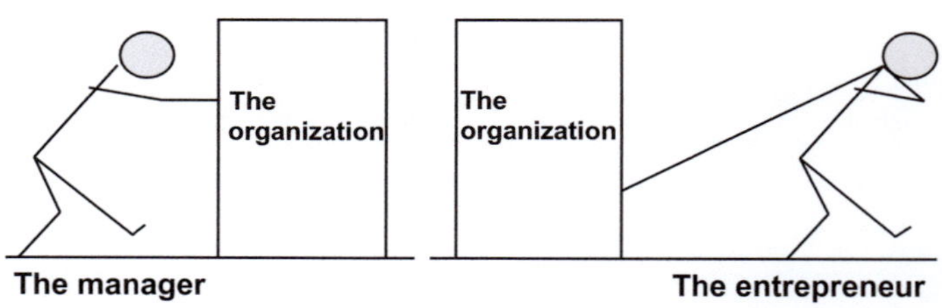

Fig. 3.2 The entrepreneur needs to be in front of her/his own organization, showing the road to take, while the manager ensures that the organization follows orders

Table 3.2 Some differences between the entrepreneurial and managerial view (based on www.effectuation.org/FAQ.htm#Process)

	Entrepreneurial view	Managerial view
Issue	**Effectuation**	**Causation**
Givens	**Means** are given	**Goals** are given
Decision agenda	what **effects** can I create given these means	what **means** ought I accumulate to achieve these goals
Underlying logics	To the extent that we can **control** the future we do not need to predict it	To the extent that we can **predict** the future, we can control it
Basis for commitment	**Can**: do what you are able to do	**Should**: commit based on rational choices from analysis & maximization
Predisposition toward risk	**Affordable loss**: control downside—avoid fatal risk, pursue zero resources to market	**Expected return**: pursue (risk adjusted) maximum predicted opportunity
Attitude toward outside firms	**Partnership**: build your market together with customers, suppliers and even prospective competitors	**Competition**: constrain task relationships with customers and suppliers to just what is necessary

The extreme view of the entrepreneur compared to the extreme view of a manager is shown in Table 3.2.

Effectual reasoning (the first term in Table 3.2) is based on the logic that "To the extent that we can control the future, we do not need to predict it". Being in an unpredictable market means that the market can be shaped through own decisions and actions working in conjunction with pre-committed *stakeholders* (i.e., a person, group, organization, member or system who affects or can be affected by an organization's actions) and customers-partners. Causal reasoning is based on the logic that to the extent that we can predict the future, we can control it.

The most successful entrepreneurs are capable of both effectual and causation reasoning and use both modes well, but they prefer effectual reasoning over causal reasoning in the early stages of a new venture, and arguably, most entrepreneurs seem not to transition well into latter stages requiring more causal reasoning (Sarasvathy 2001). Most often, they start very small with the means that are closest at hand, and move almost directly into action without elaborate planning (Sarasvathy 2006b).

In general, whereas the causal entrepreneur seeks to find a big market and then strives to capture a large piece of that big market, the effectual entrepreneur seeks to own entire or large pieces of small markets that she or he stitches together into a large market down the road (Sarasvathy 2006). Unfortunately, intrapreneurs and entrepreneurs backed by venture capital are often already pushed to behave as causal entrepreneurs from the startup of the business.

In contrast to the situation for managers, the entrepreneur has to constantly work in a field of tension between different extremes, and to steer the business through hidden reefs and skerries (and perhaps also mines) that are to be found in the waters that she/he must navigate. The entrepreneur must, for example, accept risks as being an integral part of the business, but to simultaneously actively work to minimize them as quickly as possible. They must also be able to cooperate without losing control of the business in order to successfully manage dualism in business demands creativity at many levels, with a "both/and" attitude rather than an "either/or" attitude (Johannisson 1996).

Not all entrepreneurs manage to keep up with the business in the increasingly more formalized reality that is demanded when the company grows, but instead, start again with a new innovative business.

3.4 The Qualities of an Entrepreneur

In order for the entrepreneur to be able to succeed in the difficult work of making her/his innovative project become a successful company, or to redefine the company or a department of a company, a number of qualities are required. The most important qualities are:

- strong orientation towards performance and results
- great decision-taking power (dare to take risks)
- good ability to build personal contact networks and to select staff
- good business sense
- great "survival ability"

Successful entrepreneurs have a strong ability to survive, and are experts at extricating themselves from difficult situations. They truly have learnt to live on the edge of chaos and to benefit from the possibilities chaos situations provide. The fact is that the big steps in human development sometimes have been initiated by those who balance on the boarders or sometimes have stepped over the boarders of existing laws, rules, or traditions.

The quality of being result-orientated means that in all imaginable ways, the entrepreneur will pull the product forward so that it becomes a successfully implemented in the market without much regard for her or his own person. The entrepreneur will fight or try to bypass everything that disturbs her/his potential to succeed with the product. This sometimes makes the entrepreneur a difficult person to handle for bureaucrats, administrators and controllers. The true entrepreneur will only accept administrative systems and procedures if they are favorable to her or his own goals.

In order to succeed, the entrepreneur must have considerable self-confidence in her/his own ability to manage the task in question. Many will doubt the entrepreneur's chances of succeeding with the innovative project as planned or wished. Self-confidence does not, however, imply that the entrepreneur may not listen to other opinions or evaluate them in a critical way in relation to her or his own point of view. Such considerations can increase the tempo of the business and can contribute to reducing the risk of failure.

The overall knowledge and surroundings of a new (radical) product are often small, particularly during the first phases of an innovative business. This implies that decisions must often be taken on loose grounds, which, in turn, can mean that the entrepreneur must be one step ahead of her/his organization in order to stake out the way forward. This does not necessarily lead in the same direction as first intended. Thereby, chaotic situations can arise because something that can go wrong also has the tendency to do so. An accident often does not occur alone. The entrepreneur must, therefore, keep a cool head and be able to think clearly and act sensibly, even if the whole surrounding world sometimes seems to be falling to pieces. In other words, the entrepreneur must possess considerable perseverance and resistance to stress. Sensible decisions do not necessarily have to be perceived by others as being sensible decisions! Unfortunately, the full effects of a decision cannot be seen until afterwards.

As a part of the learning process, the results of each new try must be continually analyzed in depth so that further knowledge may be gained about, for example, the product and the market. In principle, no new tries should be started until previous ones have been carefully analyzed. Maximum learning requires that the entrepreneur has good analytical abilities and that she/he be unpretentious and creative in finding new solutions. Subsequently, it is very important to be able to see a failure as a part of the learning process. Excuses and defense of the way of running a failed business do not lead to new knowledge. Shortcomings that continue to survive become increasingly costly to correct as time goes by!

Generally, a successful entrepreneur is prepared to take calculated risks in order to succeed, and tests her or his ideas in small steps so as to reduce risk exposure. The entrepreneur does not willingly take chances, except possibly in extremely tight situations when there is everything to gain and little to lose. Thus, the decision-making force is a vital necessity for the management of every innovative business.

The entrepreneur must dare to take decisions even if they are occasionally wrong or not totally appropriate. This is part of the learning process. On the other hand, a shortfall in ability to take decisions will have a paralyzing effect on the business, which, in turn, will imply hidden costs that can add up to large amounts.

The entrepreneur must not be prestigious or afraid of criticism when she/he must make a change as a result of new knowledge being available that enables a better decision to be taken. The entrepreneur must often revalue her/his opinions and be prepared to make modifications in which the small steps in the orientation of the vision of the business to be are important. The entrepreneur ought not to be prevented by her/his surroundings from making such modifications when such are required. Relatively abrupt changes, especially in the initial stages of the innovative process, can be taken as being distracting and to be an example of nervous action. Abrupt changes, however, are often a sign that the entrepreneur quickly adapts to changing situations and are therefore generally very positive.

The entrepreneur who cannot see the amusing side of various situations, no matter how bizarre they may be, will have difficulty in succeeding in her/his work. Humor is an extremely valuable asset both internally within the business and externally, for example, when selling ideas and products. Humor is also a mental safety valve for the entrepreneur and her/his surroundings when in difficult situations. Panic situations with dangerous panic decisions as a consequence can thereby be avoided. To be able to laugh a few times each day improves the atmosphere and is also considered as having a certain medical value.

One of the characteristics of an entrepreneur is also the ability to negotiate and an excellent knowledge of the subject is a considerable asset. However, the entrepreneur does not necessarily need to be knowledgeable about all of the trade or industry in question. However, with project management of technical innovations, it is an important asset for the entrepreneur to have a good technical understanding, because she/he will otherwise become very dependent on other people's opinions. Without sufficient knowledge of her/his own, the entrepreneur will not be able to make the quick, good decisions that can be necessary for the business to be successful. In short, the entrepreneur precedes her/his organization and pulls it forward towards the goal and vision on the path that she/he finds accessible and navigable. The guiding principle should be the Pareto Principle or 80/20 Principle (Koch 1998), which means accepting incomplete decisions. The manager, on the other hand, pushes her/his organization forward on a wide and well-prepared path following the principle of doing it right from the start (the 100/0% principle), which (very) seldom works well. By going first, the entrepreneur can quickly act and react in order to reduce the risk that the whole organization will go wrong. If the entrepreneur loses touch with her/his organization by going too far forward, and does not have united employees as a link between her/him and the large organization, the business will most probably fail.

3.5 Entrepreneurship from Early to Late Stages

Entrepreneurship means how a good entrepreneur should behave in different situations in order to ensure that a new business she/he is forming will grow to become an economically sustainable business and company in the long run, and entrepreneurship can be taught.

From the start of an innovative business to the time that it has become firmly established, the period of development transforms from constructive vagueness through subjective insight to objective control. The transition from intuitive mental pictures to analytical forms of description entails increases in the degree of formalization (Nyström 1996). At the same time, communicability and the prerequisites for collective decision-making increase. The demands that are put on the entrepreneur from the early stages in the development through to the later stages can be seen in Table 3.3.

Table 3.3 The demands on the entrepreneur change from the early stages of the development of a business through to the later stages (Nyström 1996)

Early stages	Late stages
Association richness	Focusing
Intellectual flexibility	Targeting
Tolerance of diversity—duty ("both–and")	Singularity ("either–or")
Visual thinking	Formalism
Rethinking ability	Consequence

Table 3.3 emphasizes the importance of the entrepreneur in the early stages of an innovative business being able to handle ambiguity, subjective insecurity and paradoxes. If the entrepreneur can manage these, and the business, whether it is a private, a public or a non-profit unit grows, the demands for focusing and formalizing will increase and the entrepreneur will consequently move from tangible action and vision, to planning in pace with the growth of the business.

Basic probability theory shows that by merely being willing to fail once or twice, an entrepreneur can increase the probability of his or her success over any given success rate. It is important to create an environment where people know it is okay to fail. That way, they will try a lot more. However, it is only okay to make unique mistakes, not to repeat them. A problem especially for the public sector is that, in principle, mistakes are not allowed at all (Engholm Jensen et al. 2008) so the important learning principle of "trial-and-error" is hampered.

3.6 Leadership

According to the Project Management Institute (PMBOK® 2000, p 25), "leading … involves the ability to 'get things done.' It requires an understanding of both the formal and informal structures of all the organizations involved….". Politics in an organization that can also greatly influence the project as well as the cultural situation, as was discussed earlier.

The activity of leading is called leadership and to lead basically means to show the way for others to reach a goal. Every activity done by more than one person will only function well if there is some kind of leadership for the activity. If no formal leader is appointed, self-organization will occur and the leadership role can jump between team members depending on situation and time.

We can define three different styles of leadership.

- *Laissez-faire (free) leadership*: This type of leadership means that the team members have to organize themselves to be able to do the work. After some time, an informal leader will generally be appointed by the team. However, if an informal leader does not appear, and if the team members cannot or will not cooperate with each other, free leadership can lead to anarchy. Experience indicates that flat line and matrix organizations (organizations with few management layers) in large organizations will lead to Laissez-faire leadership if the managers on each level have more than about ten subordinates. This is especially the case if the leader is "weak" and has problems making decisions. In such cases, and if the team members fit well together and no one takes personal advantage of the situation, a "we-feeling"will become dominant. On the other hand, an "I-feeling "becomes dominant if an informal leader with an authoritarian leadership emerges in the group.

- *Authoritarian leadership*: The manager/leader decides everything on her/his own without first hearing the views of the subordinates. As orders go down and information goes up, a (vertical) line organization chart can be sketched to show this reality. The more authoritarian the leader gets, the less the subordinates will be allowed to have direct contact with each other. The ultimate and optimal situation for the authoritarian leader is to have robot-like people who will not cause emotional problems and that will not do things in ways other than what the leader has thought of and ordered. The security of the team members is dependent on the degree to which the leader needs to be and is satisfied. To some extent, the leader looks down on the subordinates and, in bad cases, can try to increase the respect for her/himself by turning one (or more) of the team members into a scapegoat or an object of abuse or ridicule. An introverted "I-feeling" becomes dominant in the hierarchical structure formed by this type of leadership.
- *Group-oriented leadership*: In this case, the leader has a softer/human touch, acting more like a guiding parent for the team. Such leadership does not mean that the leader has abdicated the position as leader, but has chosen another way to get things done as she/he is convinced that the result will be reached faster and better with less energy-consuming conflicts if the subordinates can influence the work and are required to take responsibility for their own contributions. Therefore, the leader starts dialogues about how to solve principal problems and to choose which directions to take when crossroads and dead ends are reached. The decisions are, in general, taken in consensus and democratically. Doing so, the team members will understand why a certain decision is taken and will thereby be positively motivated. Team members being encouraged to have cross contacts with each other support the group-oriented leadership and a "we-feeling" will develop in the team. This leadership is characterized as Nordic leadership (see Fig. 1 in Preface).

In short, the three types of leadership can be symbolized respectively by:

- Trust
- Control
- Dynamic feed-back

Trust without control is like "black box" leadership because the leader hopes that things will be as agreed upon, which they seldom turn out to be. Thus, in this case, the leader has little or no interest in looking into the box to influence the work being done there. When the leader imposes controls in advance on the work done in the box at "gates" or "tollgates", she/he will get a glimpse of the situation at these control points. Still she/he mainly has to trust the information given without really knowing the situation.

"Jumping into the box" and giving feed-back on what the leader experiences refers to a situation when she/he knows quite well what is going on. That person is far better able to guide the project to a satisfactory situation and result than for the other two alternatives. By setting realistic milestones (see below) and goals and adjusting them when needed makes the development successful. By giving continuous feed-back, the people she/he is leading will feel well. That will, in turn, result in a better job being done.

3.7 Coaching and Mentoring

On the road from the start of an innovation project to a sustainable business, in urgent situations the entrepreneur may need to give orders, although acting as a coach and a mentor is often preferable to get the team members to be engaged in an optimal way.

Coaching and mentoring are used to improve the results for individuals and teams in sports and job situations. The scientific backing is weak for "coaching" and "mentoring" although there are many connections to psychology. Generally seen, coaching seems to differ from mentoring in focusing on specific tasks or objectives, as opposed to general goals or overall development for mentoring (e.g., Garvey et al. 2014 or Renton 2009). Further, coaching is not synonymous with training, but training can be one component of a larger coaching initiative (Chakravarthy 2011). Some differences, seen from a practical view between coaching and mentoring are shown in Table 3.4.

The coach develops specific skills for the task, challenges and performance expectations of the work. The mentor should have a deep personal interest and be personally involved, as a friend who cares about her/his friends and their long term development.

In short (www.coachingandmentoring.com/Articles/mentoring.html):

A **coach** tries to direct a person to some end result. The person may choose how to get there, but the coach strategically assesses and monitors the progress and gives advice for effectiveness and efficiency. Coaching is impartial, focused on improvement in behavior.
A **mentor** is like a sounding board. They can give advice but the partners are free to pick and choose what they do. The context does not have specific performance objectives.

Another view is (www.td.org/Publications/Blogs/Human-Capital-Blog/2014/08/Mentoring-Versus-Coaching-Whats-th):

Table 3.4 Some differences between coaching and mentoring (from www.ucl.ac.uk/hr/od/coaching/differences.php)

Coaching	Mentoring
Coaches need not have first-hand experience of the coachee's line of work. The coach can be an independent external professional with expertise in coaching, or a qualified internal coach	Mentoring is customarily a planned pairing of a more skilled or experienced person (usually in the same field of work) with a less experienced
Line managers can use coaching techniques successfully in the management and development of team members	Ideally mentors have no line management relationship to the mentee
Coaches will ask 'powerful' question and not offer or give advise	Mentors will often provide direction and advice and should open organizational doors' for mentees
A number of both internal and external coaches are available with a variety of backgrounds and expertise and the services they provide tie in with the organization's objectives	Mentors can provide a neutral 'sounding board', assure total confidentiality, and have no agenda other than assisting their mentees in their development and to reach their goals
Effective coaching is intended to help you to learn rather than by "teaching". By engaging with an experienced coach, the coachee will develop insights leading to enhanced effectiveness	Mentoring involves helping mentees to develop their career skills and expertise often drawing upon the experience of the mentor in the process

Mentoring: Talking *with* a person who has identified his needs prior to entering into a mentoring relationship. The emphasis is on active listening, providing information, making suggestions, and establishing connections.
Coaching: Talking *to* a person, identifying what he needs, and developing an action plan. The emphasis is on instruction, assessing, and monitoring.

Mentoring seems to happen through story telling from other people's experiences and by relating anecdotes of the mentor. Thus, mentoring can be abstract. It is a power free, two-way mutually beneficial learning situation where the mentor provides advice, shares knowledge and experiences, and teaches using a low pressure, self-discovery approach. Mentors share their own experiences and skills so that the mentor is both a source of information/knowledge and a Socratic questioner (Starcevich 2016). Mentoring focuses on the individual and the conversation transcends more broadly into the general work. This means that the interaction can be more philosophical, more focused on attitudes and behaviors than on specific skills (Chakravarthy 2011). That is why mentoring should be valuable for student professional training in higher education.

Entrepreneurs train and coach the people who are engaged in their innovation projects. To do mentorship demands life experience that grows with age. The comets used in the planetary organization (see the next chapter) should act as mentors, meaning that they must be life experienced.

3.8 Building an Identity

For any project team to be efficient, the feeling of identity is important. The entrepreneur has a special responsibility to see that each team member develops a good feeling of identity towards the project.

In a broader perspective, cultural identity is the feeling of identity of a group or culture, or of an individual as far as she/he is influenced by her/his belonging to a group or culture. Corporate Identity (CI) is the reality and uniqueness of an organization, which is integrally related to its external and internal image and reputation through corporate communication. CI is commonly viewed as being composed of three parts:

- *Corporate Design*—trade marks, logos, uniforms
- *Corporate Communication*—web pages, commercials, public relations, events, and other information
- *Corporate Behavior*—history, ethical and moral codes, internal values, and norms

In any activity where more than one individual takes part, the common sharing of the identity is crucial for the wellbeing and outcome of the activity (see Fig. 3.3).

A common, accepted identity also helps to attract external interest, meaning, for example, that other people will be involved in the activities—working there, investing money, or testing equipment. Another positive consequence of a unified identity is that engagement fosters full utilization of knowledge and skills (Hackman 2002). Increased identity will also help to increase the quality of the work being done, as well as increase the self-confidence and the efficiency in the work. However, having people in the group that do not want to share the identity creates a problem, which is why it is sometimes better to separate them from the group.

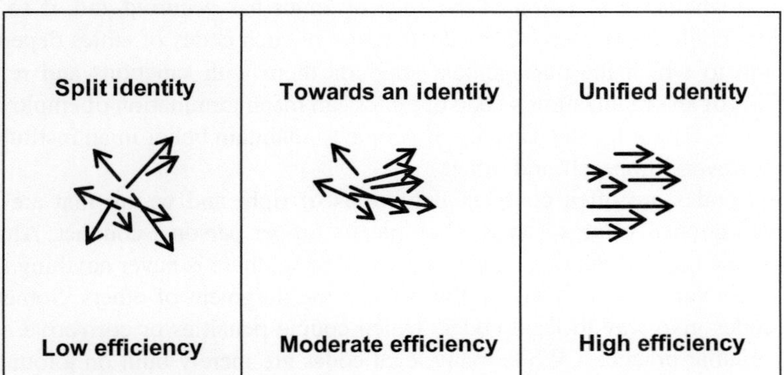

Fig. 3.3 When a split identity in an organization is transformed into a unified identity, the efficiency increases as well as the wellbeing and other positive effects

When traditional manning principles are used in larger organizations—when all team members are assembled simultaneously—for the intrapreneur, the start will often be a split identity as the people joining will have different views of life and different experiences. The first attempt for the inprapreneur to get a unified identity in that case can start at a so called *kick-off meeting,* followed by continuous work until the unified identity situation is reached. When successive manning is used, the project leader can form the team with the "right" identity as an entrance demand. Thus, the behavioral part of the identity is strongly connected with ethics and morale.

3.9 Ethical and Moral Codes

Strictly speaking, ethics are beliefs, while morale is the behavior of a person.

The word "ethics", comes from the Ancient Greek "ethikos", meaning "theory of living", and it is a major branch of philosophy.

Ethics deals with the study of values and customs of a person or group and covers the analysis and employment of concepts, such as right, wrong, good, evil, and responsibility. It can be divided in the three primary areas:

- meta-ethics (the study of the concept of ethics)
- normative ethics (the study of how to determine ethical values)
- applied ethics (the study of the use of ethical values)

An ethical code is a code of professional responsibility. A code of ethics for a company or an organization is often a formal statement of the organization's values on certain ethical and social issues. Some set out general principles about an organization's beliefs on matters such as quality, employees and the environment. Others set out the procedures to be used in specific ethical situations, such as conflicts of interest or the acceptance of gifts, and they delineate the procedures to determine whether a violation of the code of ethics has occurred and, if so, what remedies should be imposed. The effectiveness of such codes of ethics depends on the extent to which the management supports them with sanctions and rewards. Violations of an organization's code of ethics can mean termination of employment.

Morale is a term for the capacity of people to maintain belief in an institution or a goal, or even in oneself and others.

Moral codes are often complex definitions of right and wrong that are based upon well-defined value systems. They dictate proper personal conduct. Although some people might think that a moral code is simple, there is never anything simple about one's values, ethics, or, for that matter, the judgment of others. Sometimes, moral codes give way to legal codes, which couple penalties or corrective actions with particular practices. While many legal codes are merely built on a foundation of cultural moral codes, they are often one and the same.

In general, for larger organizations there are moral and ethical codes for all employees and associated people (as consultants) to follow in the daily work. For independent entrepreneurs and renovators, it is their responsibility to set the moral codes and to ensure that they are implemented in the innovation project and sub-projects.

Note that there are also ethical laws and regulations for certain occupations, such as doctors, or scientists.

3.10 Dealing with Stress and Conflicts

Stress and conflicts occur in all projects, but especially in innovation projects, because new behavior and changes cause stress situations. Some stress has the advantage of making us function and perform at our best (Rasulzada 2007). Note, however, that we all tend to perform at the maximum close to deadlines, which, for instance, is shortly before the presentation of a new product at a fair. Thus, efficiency and performance increases with increased stress up to the point that the person's capacity to manage and cope is exceeded, at which point the negative stress leads to communication difficulties and irritation. In more severe cases, it can lead to headaches, sleeping problems, increased blood pressure, gastric ulcers and eventually burn-out.

Stress levels vary largely from person to person and have different manifestations. A project leader who has a tendency to easily get stressed will also stress her/his subordinates, which increases the risk of mistakes and conflict situations. A good project leader is therefore like a duck. It appears that the duck flows calmly on the water, but under the surface, it paddles furiously.

When some people get stressed, they can have problems making decisions and will, in severe cases, such as burn-out, be paralyzed and even apathetic. Others will make decisions too quickly, without having collected the necessary information on which to make them. Some will just be 'not nice' and even aggressive when the perceived stress level gets too high.

A particularly stressful situation is when one is not allowed to use one's own knowledge and experiences, even though the tasks to be done may require precisely that knowledge and experience. To be given under-qualified tasks or to be unfairly judged and treated, or to be mobbed (bullied) causes hurt and bad feelings. Creative people sometimes claim they can generate force out of such treatment, while others can be totally broken. People experiencing high levels of positive effects, such as enthusiasm, happiness, and interest tend to be positively engaged and feel good about the activities in which they are involved. People feeling high levels of negative effects, such as boredom and depression, tend to become disengaged to some degree from the world around them (George 1996).

Farida Rasulzada (2007) found that organizational creativity and innovation leads to better well-being and that a creative organizational climate leads to less stress. She also found that gender was more important than education with regard to stress and workload. Less stressful situations occur when different opinions exist on how something should be done and who is going to do it. Different choices to be made can cause conflict, leading to stressful situations. This is also the case when someone feels she/he has to work more than other team members who do not take their responsibilities seriously. Stress situations can be foreseen and somewhat neutralized by ensuring that the personal chemistry functions well between the team members at the recruitment stage. The fewer that are in a team or a sub-team, the more important the personal chemistry will be in helping avoid unnecessary stress and conflict situations.

Heavy workloads over long periods of time can be a health hazard, although different persons have different tolerance levels where fatigue is concerned. It is important that the project leader and the sub-project leaders keep an eye on how the different team members feel, which is accomplished by the use of Management by Walking Around. Using that management principle also helps to avoid misunderstandings in the organization as the leaders can see how people work and act. Being frequently present where the work is actually taking place also means that one gets a feeling of whether the 'climate' is good or not. In that respect, the Planetary Organization is also better than a line organization (see the next chapter).

When conflict occurs, it is important to strive towards win-win situations, holding on to what is important and sacrificing what is not. To reach such a situation requires that the project leader intuits outspoken and unspoken wishes. She/he must also arrange for dialogue and has to be creative and humble in order to find functional solutions. Note, however, that discussions over the Internet easily lead to more complicated situations that later must be solved with dialogues.

To lower the risk of stress levels that are too high, everybody in a project would be well advised to follow some good advice given by diverse experts on the problem:

- Arrange for spare time by listening to music, reading a book or simply going to see a movie, whatever helps you relax.
- When you feel you are too busy try to calm down, for example by sitting down and thinking over the situation.
- Set priorities and take away 'musts'.
- Set up limits for what you can do.
- Don't run away from conflicts but soon in a creative way try to solve them.
- Speak with someone when you feel stressed.
- Be physically active as that is the best way for the body to release stress. For example, walk half an hour every day.
- Do relaxation exercises.

3.11 Burn Out Problems

The term 'burn out', whether occupational burnout or job burn out, has been around since the 1990s (Jackson et al. 1986). It was noticed especially in young software programmers and means a complex and severe situation for those concerned. When a person has been burnt out, she/he will have a very long period, perhaps years, of recovery to get back to normal again.

The symptoms of an approaching burn out situation are, according to medical experts, increasing difficulties in sleeping that can escalate to difficulties in eating. The stress hormone levels get more and more affected so that gradually the muscles get more and more tense, reducing the flow of oxygen to the cells. This causes aches and pains of different kinds, such as lumbago, stomach and intestine trouble, and heart attacks.

When waste is stored in the cells, the higher functions in the body are affected and the person starts to lose memory, and concentration difficulties become more noticeable. She/he can swing between crying and being angry, and will lose engagement, empathy and the ability to love. Judgment disappears and her/his movements become mechanical.

When someone has been burnt out, all the warning signals have been ignored in a serious, inhuman, un-ethical and indefensible way. The project leaders are responsible for seeing that no team members reach burn-out.

A serious problem is when the entrepreneur is approaching a burn out situation. If the entrepreneur is an independent entrepreneur, family and friends have to react. If the entrepreneur is an intrapreneur or renovateur, other bodies have to react.

Chapter 4
Project Management

4.1 Projects in a Historical View

The word "project" comes from the Latin word projectum from projicere—"to throw something forwards". In turn, "pro-" (from the Greek "πρό") denotes something that precedes the action of the next part of the word jacere—"to throw" (source: Wikipedia).

The word "project" therefore originally meant "something that comes before something else is done". When the word was initially used, it referred to a plan of something, not to the act of actually carrying this plan out. Something performed in accordance with a project was called an object. Seen in this perspective, the building of the pyramids in Egypt was an early project. After WWII (World War 2), however, the meaning of a project changed to what can be called modern project theory.

Modern project theory has its roots in the product development (PD) of the first US nuclear missile, the Atlantis, from 1954 (Morris and Hought 1987). The military PD projects were big, and different planning and guiding tools were needed and developed to control the time and cost limits. Therefore, for example, the PERT—Program Evaluation & Review Task—and the CPM—Critical Path Method, which we will come back to, were developed in the early 1960s. The terms "project leader", "project organization", and "matrix organization" became profitable new fields for consultants, often with military backgrounds (Engwall 1995).

In 1997, the International Organization for Standardization (ISO) developed the standard ISO 10006, titled "Quality management—Guidelines to quality in project management". It is part of the ISO 9004 standards. As the title suggests, the focus of ISO 10006 is how to achieve an even quality in project activities (e.g., even bad quality or even good quality as two counterparts). The document stated that it is "not a guide to project management itself," which is why it contains prescriptive terms such as "action should be taken", decisions should be "formally documented", "special attention should be given to", and "particular attention should be

S. Ottosson, *Developing and Managing Innovation in a Fast Changing and Complex World*, https://doi.org/10.1007/978-3-319-94045-8_4

given to", making it less useful for practitioners. Thus, with the ISO 10006, standard quality (Q) was introduced into project philosophy.

Today, there are several associations that address project management knowledge. Two examples are the Association for Project Management (APM) and the Project Management Institute (PMI). PMI produces the Project Management Body of Management (PMBOK® Guide), in which a project is defined as: "a temporary endeavor undertaken to create a unique product or service."

Modern project theory defines a project to have three aims (e.g. Burkes 2011):

- to be unique
- to be temporary—that it has a start and an end date
- to have a limited cost budget

Projects normally have no income and differ strongly from operations/processes, which delivers the same or almost the same results over and over again.

A common graphical presentation of the transformation process from an input to an output of a project is seen in Fig. 4.1. In this process, the project is limited by a set of constraints and utilizes a set of mechanisms. The input is often a (customer) *need* or a demand from a customer or the market. It can also be based on new insights, new knowledge and wishes or wants (as shown in Fig. 1.1). The project output is the product and/or service that satisfy the input demands. The main constraints are time, cost, and performance parameters. However, these are also supplemented by other constraints, such as legal, ethical, environmental, logical, activation and indirect effects. The mechanisms that enable the transformation process are humans, knowledge and expertise, capital, tools, and techniques, and technology (APM 2000).

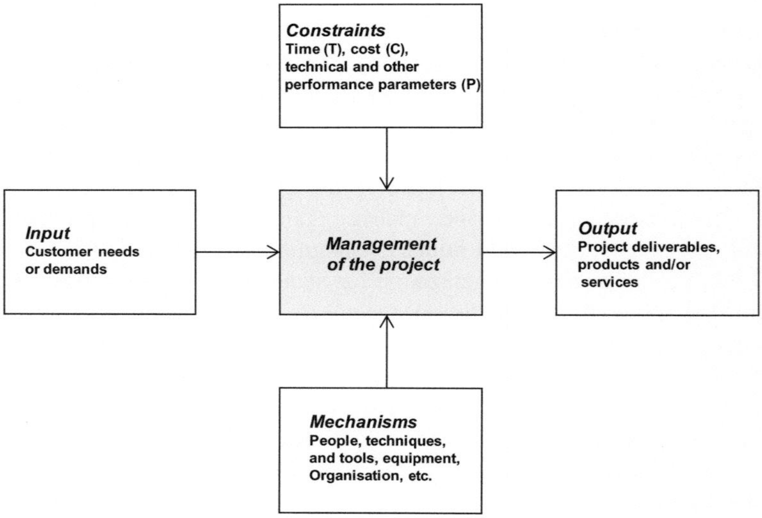

Fig. 4.1 The project management process seen from a classical point of view

When the customer and/or the board does not care much what solutions for the performance parameters are reached as long as they are in accordance with set demands on cost and time, that is often called a "black box strategy".

A special type of project is an *innovation project*, which is not treated very much in project theory. It is also unique and has a start but is gradually transformed into an ongoing activity instead of having an end date. In reality, it has no cost limits and has some income from the initial sales of the innovation when it becomes a commercial innovation. Often an innovation project—independent of if it is done in the private, the public and the non-commercial sectors—gets financial support step by step from a donor or donors, for examples, via so called "crowd funding", which we will come back to. Innovation projects need to be dynamically performed quickly in order to adjust to changing circumstances while being developed.

Innovation project theory can be seen as the most advanced project theory, as it deals with more complex situations and contexts than classical projects. This chapter focuses on project theory both from the dynamic view as well as from a classical view.

4.2 Projects and Processes

A project often is used as an appropriate form for development of new products and services for which cross professional competences are required to meet the stakeholders' needs and expectations (Burke 2011). However, an organization that wants to be sustainable in the long run must also make adjustments and re-engineering of existing products as well as developing new solutions (see Fig. 4 in Preface).

To avoid negative influences from old thinking, innovation projects can benefit from being done outside the ongoing processes. In contrast, larger adjustments and re-engineering projects, as well as improvement projects, will most probably benefit from being done as projects close to the ongoing processes. Also, minor adjustments and change projects can benefit from being done as smaller projects within the ongoing business processes.

For unique activities, such as new product development, production development, and new business development, the project organization has as mission to "draw up the map" for how to work and handle the new product and/or service. To draw up the map means to reflect on and to learn from what is experienced from different tests. In turn, that means to deal with complex (unforeseeable) situations until a stable situation is reached. For repetitive work, such as mass production of standard products, one can deal with foreseeable situations for long periods of time, which sometimes can be complicated by influences of complex situations that call for change projects to solve the situation. Therefore, for stable and easy to plan situations, a process organization symbolized by the following the map can be used.

While it is natural to integrate the result of a change project into a healthy process, it is quite tricky to know how and when to integrate an innovation process into the standard process of a company. If implemented into the daily operation, a rule of thumb is that at least 60% of the people in the new unit initially should be taken from the innovation project. According to a manager tasked with the setting up new IKEA warehouses, this is how IKEA has found the best way to set up new furniture stores—bringing in 60% of the people from other units for the sensitive start-up phase that in their case lasts for at least twelve months.

Thus, the implementation of changes and new products in an existing organization is complicated. Investigations (Beer and Nohria 2000) have shown that 70% of change initiatives were met with resistance, and as a consequence, failed. Vakola and Nikolaou (2005) showed that change often was met with a negative attitude that brought stress and insecurity among the people in the organization.

4.3 Complexity Aspects of Development Projects

More and more, we experience the world to be increasingly complex, un-ordered, and non-linear, for which the classical project management principles were not developed (Holmdahl 2007). This is especially the case when carrying out an innovation project, for which a New Product Development (NPD) project is a sub-project. The complexity comes mainly from the fact that such projects build on few known solutions. Therefore, people inside (and outside) the development process will often, especially in the beginning of the work, experience chaotic and complex situations.

As mentioned, until a stable product and market situation has been reached, it will feel like walking through a labyrinth. Thus, the developers frequently have to rely on very little and/or unreliable information when deciding on an action or solution. If that decision leads to a dead-end, which is not un-common, they will experience the frustrating situation of having to go back and try again until they have found the successful path. When dead-end situations are reached, leadership then becomes critical for the project because this will decide whether or not the project will come back on track.

However, complexity comes not only from external influence (e.g., society, competitors, customers, or users) but also from internal sources (see Fig. 4.2). Thus, the risk of failure in one or more respects, because of overdrawn costs and time, low satisfaction rate, poor performance, and burnt out, team members are dependent on many factors.

External contextual influence includes laws and regulations. Note that, to some extent, entrepreneurial work involves exploiting the rules and laws to the limit and not to setting up more barriers and limitations than those already existing. For example, within the public sector, procurement rules can severely hamper innovation if the procurement rules are hard drawn.

Most of the factors in Fig. 4.2 are 'soft' and difficult to measure and handle. Maybe that is why most research interest, so far, has been focused on tools and methods. Sometimes it seems as if people think that if the "right" tools are used, a good result of a development process is guaranteed. That is as bad as believing that having a good

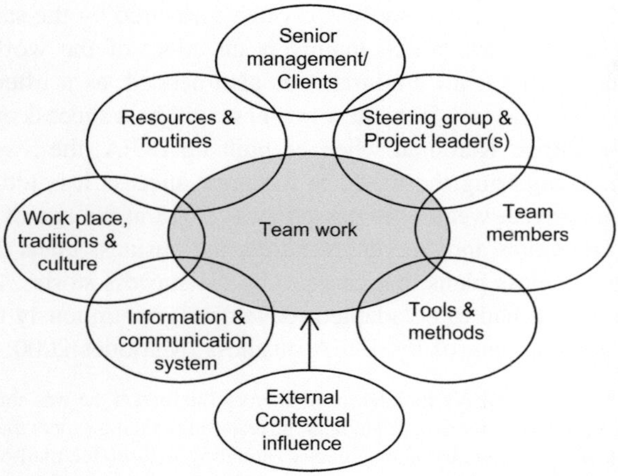

Fig. 4.2 Product development (PD) projects are often dependent on many factors that, in turn, are time dependent, making them truly complex/unforeseeable

business plan for a new product will guarantee success, if only enough (venture) capital is available. Thus, having a good business plan, a well thought business strategy and the right products will not guarantee success. (In the Lean theory it is claimed that 80% of the result is dependent on the people and 20% on the tools.)

The success of a new product development (NPD) or an innovation project can be measured in many dimensions. Quality (Q), cost (C), and time (T) are three dimensions used in classical project theory, forming what is called the corners of a project triangle. Sometimes, it is claimed that if the center of gravity were in the center of that triangle, it would constitute a good project, which it never can or should be. For instance, when a new nuclear power station is to be built, Q is the most important parameter. For the development of a new product with a short commercial lifetime, the development time (T) can be the most important parameter. For an organization with a weak financial situation, the cost (C) can be the most important parameter. If the development of a new product were completed according to plan but resulted in a product that cannot be sold or is not used much by those who have bought it, it was not a very successful development even though the limits of Q, C, and T were held.

Another weakness with the classical project theory is that Q is not a good term to measure against for PD projects as it is difficult to define. Instead, demand fulfillment, or "Performance" can be used, for which Quality (Q) only is a sub-set of performance (P). The importance of P, C, and T will differ for need-based PD, want-based PD, and wish-based PD. In reality, it is very difficult to get a correct picture of C, because for micro and small companies, much of the work is often done at no labor cost and using test pieces free of cost. Also, in medium and large companies, many projects profit from other projects and therefore do not show the correct numbers to the product boards. Many of the most successful innovative projects are known to have been financed this way.

For innovation development, success is often measured by the sales growth and profitability of the product, which mainly is an effect of the work done by the entrepreneur and her/his team. Endurance is also needed, as it often takes four to five years to be relatively sure if the new business will be a success or not. That was the case for Mr. Ingvar Kamprad when he built up IKEA, the Swedish furniture company that is conquering the world to the great success it is today, and giving him an enormous private wealth (he passed away 2018 at the age of 92). The same is true for Mr. Bill Gates and Microsoft. Surely the business plans they started out with are not the business plans that came to be the success stories. An example of how the business idea had to be adapted (developed) continuously to meet market realities is the development of the IKEA business (Markides 2000, p. 150):

> When Ingvar Kamprad, IKEA's founder, tried to crack the market, he was shut out at every turn. Barred from selling directly at trade fairs, he resorted to taking orders there. When that was forbidden, he contacted customers directly (initiating a profitable mail-order business, which required that the furniture be easy to ship). When Swedish manufacturers refused his business, Kamprad sourced from Poland, getting even better prices than before. Locked out of traditional outlets, Kamprad converted a factory into a warehouse and showroom, where explanatory tags, self-service, a colorful catalog, and the lure of instant availability — thanks to on-site stocking — were deliberately distinctive. In every instance, the strategy was driven as much by necessity as it was choice. In hindsight, IKEA's positioning is indeed brilliant and is indeed a source of real and sustainable differentiation. This position, however, was as much a consequence of adaptability as it was of strategy. It was persistence—and experimentation under the strict discipline imposed by constrained resources—that allowed IKEA to build its furniture franchise.

4.4 Three Types of NPD Projects

As described in Chap. 1, there are many backgrounds to starting a new product development (NPD) project. The most well-known start condition is to start with a need, which belongs to the group Challenge (see Fig. 1.1). In that group, there is also a want or a wish. Table 4.1 shows the main differences between the three challenge starting conditions for product development.

A *need* is normally concrete and existent, while a *wish* is distant and at a high dream level, as is shown in Fig. 4.3.

For *need-based* product development, Time-to-Market is an extremely important factor, as a need is there to be satisfied. The same need is probably being experienced by many competing actors in the market. The price of the product is also critical, as many similar solutions will be in competition, meaning that the cost to develop the product is critical for its commercial success. The performance demands are high, but must be balanced against the total price that can be accepted by the market.

Also, for *want-based* product development, Time-to-Market is an important factor, as the want can soon become a need. The want is probably being experienced simultaneously by one or more competing actors in the market. However, the

Table 4.1 Three types of backgrounds for NPD projects causing different circumstances for the project work

PD background	Characteristics	PD target	Planning	Stable conditions	Innovation type
Need	Knowledge and solutions exist to reuse for an existing need	Fixed	Fulfill plan	Yes	Product adjustment
Want	Knowledge and solutions are not enough to provide a want	Moving	Adapt to the situation	Partly	Incremental
Wish	Important knowledge and solutions do not exist	Vision	Create, make and test	No	Radical

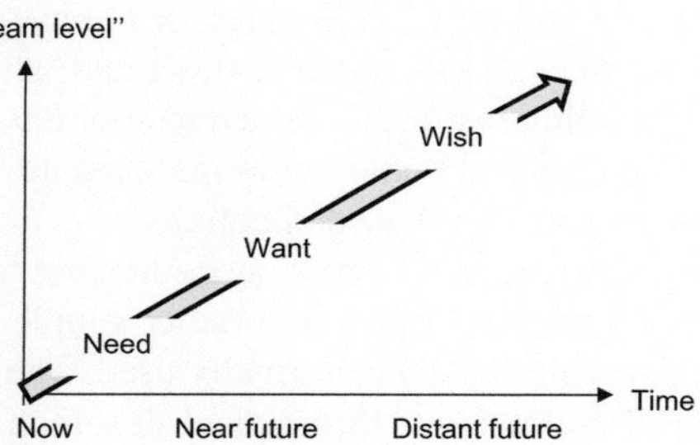

Fig. 4.3 A wish is at a high dream level that might be satisfied in the distant future, while a need is concrete and already existent

price for the product is not critical, but has to be set by comparing similar solutions, meaning that the cost to develop the product is of importance for its economic success. The performance demands are high but must be balanced against the total customer price.

Finally, for *wish-based* product development, Time-to-Market is often not an important factor, and the price of the product, generally speaking, cannot be set by doing market research. Therefore, the cost of the development is, in general, based on resources available, and the performance demands are high or difficult to reach as useful solutions are few or non-existent.

When taking on the mission to develop a new product, the "right" solution can either be what was the opinion when the decision to start the development was taken or what is right at the delivery date of the development. The first alternative is easily measurable, as the development means fulfilling demands clearly stated when the development started. The second alternative is more diffuse, as it is based on changes in the society and the market that were not known beforehand.

These two opposing views have a tremendous impact on the development process. The first alternative, here called the classical way, based on satisfying a need, means making final decisions as early as possible. The alternative way, the dynamic way, means making early preliminary decisions and making final decisions as late as possible to make it possible to take in changes and new possibilities in the ongoing development.

The first view works well for *need-based* development when all demands are clear and the time horizon to delivery is short. To make the development efficient, the collecting of as much information as possible at a detail level before the development starts is needed. Then, finding usable solutions via investigating and valuing other solutions—Benchmarking—is a way to quickly find solutions to purchase and/or to work further.

When the demands are not that clear and the time horizon is long, one gets more in the direction of a *want-based* development. To find out wants for a near future, as well as to create *wishes* for a more distant future, one way is conducting unstructured interviews and dialogues with people in the market. Other ways of finding wants and wishes are to encourage people to express their views, for example, on web pages, to study trends and research findings, or to have creative meetings. Now the start is to find a useful principle on an abstract and wholeness level, and from there, to create and test solutions until a concrete solution on the detailed level is reached. For a want-based start, some solutions exist to make the development easier and faster than for the wish-based situation.

While developing a new product from a need, the circumstances are, in general, quite well-known, which means a quick and rather simple development for which Computer Aided Engineering (CAE) is normally used. The opposite is the case for want- and wish-based development. In practical development work, many disappointments and re-takes are often needed before CAE can be used to finish the development. Figure 4.4 shows this. BAD-PAD-MAD in the Figure stands for Brain, Pencil and Model Aided Design. We will come back to these abbreviations.

In the increasingly complex society we are living in, things can change drastically with short or no or diffuse warnings in advance. Therefore, the classical

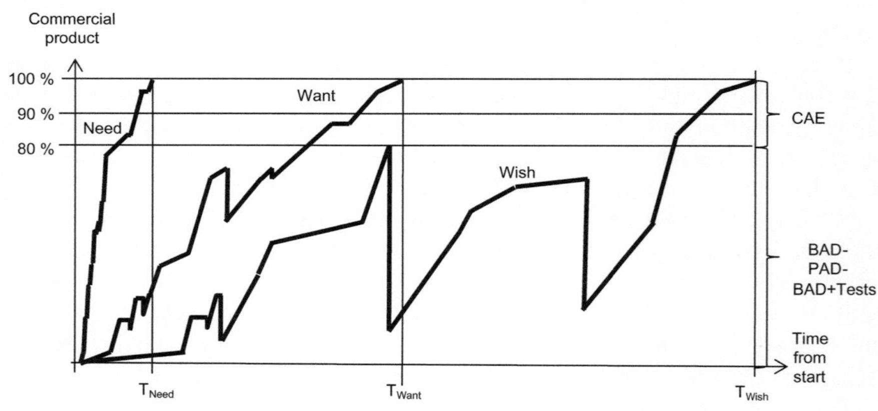

Fig. 4.4 The time from product development start of need-, want- and wish-based development to when a commercial product is ready depends on which start conditions there are. Also the shapes of the curves vary considerably

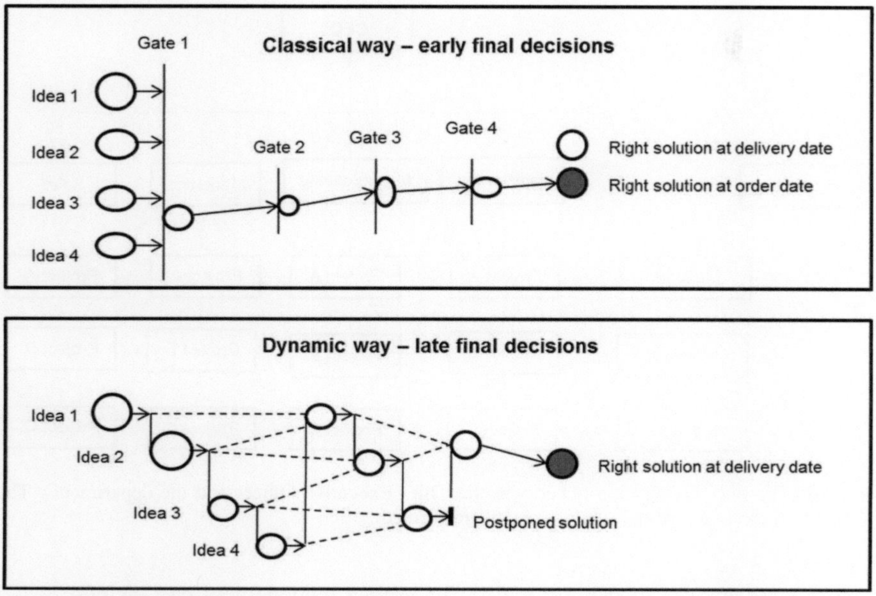

Fig. 4.5 In classical development, early final decisions are used while late final decisions are used for dynamic development

principle of final decisions that are taken early may lead to delivered solutions that are less good or wrong at the delivery date (see Fig. 4.5). The dotted lines in the Figure mean that the knowledge gained will eventually be used later in the further development of a solution.

Thus, for the dynamic situation, only a few steps backward will possibly be the situation, instead of starting the whole development over again, and this leads to faster development. This, in turn, means a great operational/business advantage and a less costly development processes. An indication of this is the leading position Toyota has in the automotive business. Toyota, which is the master of the Lean principles, has also shown it to be a less risky to work the dynamic way compared to the classical way (e.g. Kennedy 2003; Ward 2007; Kennedy et al. 2008).

4.5 Organizational Forms

A well-known form of organization is the line organization, which is used for almost all types of organizations, from military organizations, to governmental, non-profit, and private companies in most of the world. As mentioned, for project work, matrix organizations are often used. A matrix organization is two-dimensional (2D), meaning that, in principle, there are two bosses for every team member (see Fig. 4.6).

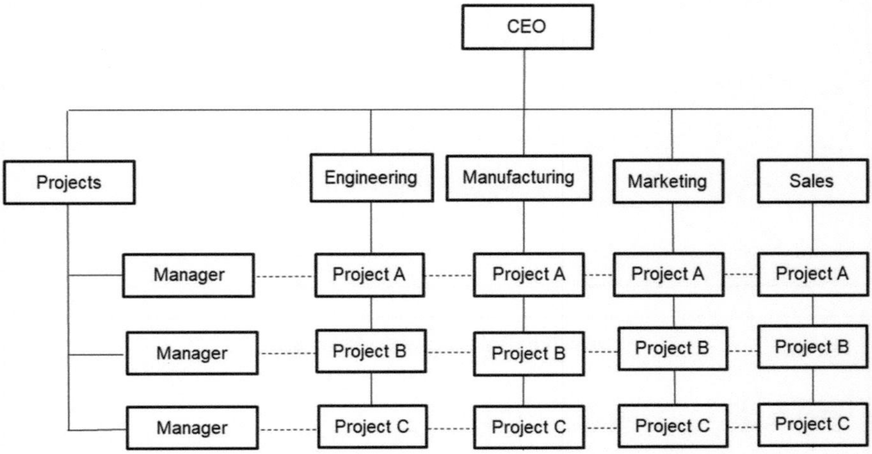

Fig. 4.6 The line organization is between the Chief Executive Officer and the departments. The matrix is in the case created when projects are formed

Some advantages of using a matrix for project management can include:

- individuals can be chosen according to the needs of the project without employing them,
- the project team will be able to view problems in different ways as the team members have been brought together in a new environment.

However, there are disadvantages for traditional project management organization models, such as:

- a conflict of loyalty can arise between line managers and project managers over the allocation of resources,
- projects can be difficult to monitor if teams have a lot of independence,
- costs will increase when more managers are used than for the line organization.

To overcome the difficulties with the line organization pattern for both types of organizations, models building on the principles of self-organization have been proposed and tested since early 1970. One of the pioneers was Sandvik AB in Sweden (Handelshögskolan 2014). Continuing in 1987, when the CEO, Percy Barnevik, moved to ABB, the Swedish branch of ABB formed a large number of self-organizing groups as a part of an effort to flatten the organization (Barnevik 2011). Today this type of management practice often is called "Circular Management", also known as "Holacracy" or "Sociocracy". The term circular implies that power and leadership are distributed throughout the organization, while maintaining an unambiguous hierarchy, conceived as a sequence of abstraction rather than authority levels (Romme 2016). This type of management practice implies that people take on roles as needed, rather than anyone becoming exclusively and (almost) permanently assigned to a managerial or any other role. More

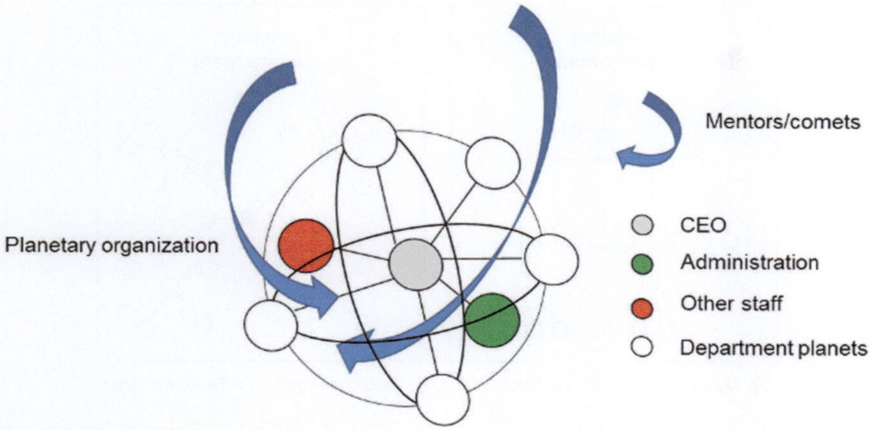

Fig. 4.7 Exercising dynamic leadership means to be present in the middle of the activities allowing a controlled freedom in the team. In this case the situation for a small- or medium-sized enterprise (SME) is pictured

information about sociocratic and holacratic approaches, as well as to circular management principles, can be found, for example, in (Robertson 2015a; Romme and Endenburg 2006; Romme 2016).

A third organization model that can be seen as a combination of the line organization and the self-organization is called "Planetary Organization" (Ottosson 1998). It was designed to take care of both vertical and horizontal communication and feed-back, as is shown in Fig. 4.7. The leader is in the center of the planetary organization, giving energy to the other "planets". There are also "comets" which move freely in the planetary organization, which we will come back to. The planetary organization has similarities with what is called "Network Organizations". However, they are more informal in their structures and are not thought to be something to organize an organization after.

Creative organizations should benefit from working out a model that would allow them to maintain a balance between non-hierarchical and hierarchical interactions, associated with equilibration and disequilibrium, or hierarchy, respectively (Shkliarevsky 2017). The Planetary Organization seems to meet that demand well.

4.6 Outsider or Insider Management?

The Line Organization has, as its base, an authoritarian leadership. The softest management is at hand when the manager only occasionally meets with her/his subordinates, which means an outsider management situation. Only at these meetings do the team members feel pressurized. The other extreme is when the manager is at the center of activities most of the time, exercising insider

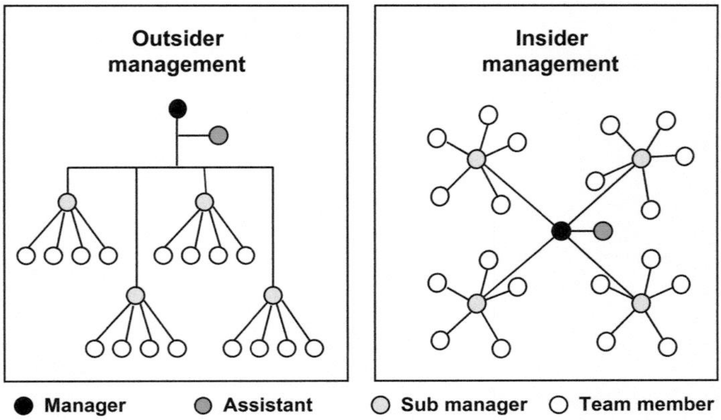

Fig. 4.8 The traditional Line Organization with an outsider management position means a softer management than an insider management position

management. This position creates more pressure on the team as the subordinates cannot easily hide realities from the manager. Figure 4.8 shows the two opposite management situations.

The Laissez-faire situation means that no formal leader/manager is appointed (see Fig. 4.9). A similar situation will also occur in a Flat Line Organization with few middle managers. The flatter a Line Organization becomes, the more dominant the Laissez-faire leadership will be. This is the result of the leader having few opportunities to maintain close contact with all her/his subordinates. The more subordinates a leader or manager gets in a "flattening" process, the more natural it also becomes for her/him to change over to an authoritarian leadership.

At the same time, the possibilities increase that one or more members of the group will step forward to become *informal leaders* in the absence of a consistent manager/subordinate contact. The informal leaders can, in the worst cases, be very dictatorial. This is because their mental power over the group will be greater than that which a formal leader could gradually build up, for example, using punishment

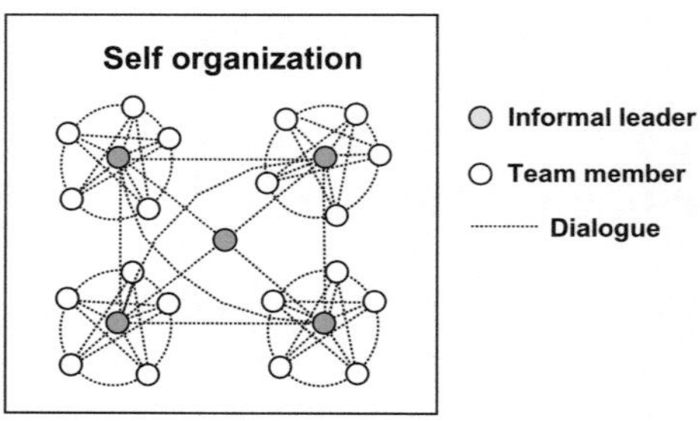

Fig. 4.9 When no formal manager (leader) is appointed, self-organization will occur, meaning dialogues take place between the team members in order to solve problems efficiently

Table 4.2 Attributes of two different organizational structures (Imperato and Harari 1996)

	Hierarchical organization	Complex adaptive organization
Organizational structure	Top-down (line)	Networked
Decision-making process	Command-and-control	Consensual
Team operation process	Appointed, managed	Self-selected, Self-directed
Leadership selection process	Dictatorial, self-aggrandizing, condescending	Unassuming, demonstrative, responsive
Leadership style	Imposed	Self-selected
Employment model	Up or out	Project to project
What gets rewarded	Past performance	Potential value of skills, experiences, relationships
Who makes enterprise decisions	Disconnected 'Top'	Small, improvisational 'Centre'
Key advantage	Efficient	Flexible

to create fear and insecurity among the team members. From psychology, it is well-known that peer pressure is much more powerful than the concept of a boss (Gladwell 2000, p. 186).

The self-organized organization in Fig. 4.9 is, in fact, a complex adaptive organization opposed to a hierarchical organization (see Table 4.2). A positive dynamic leadership uses the best of the both organizations, forming what can be called a "Dynamic Organization".

When performing team work, as well as all types of work together with other people, dynamic leadership is preferable as it is situational. This means that in critical situations, when a negative chaos situation occurs or is emerging, occasional authoritarian leadership may be needed to avoid chaos or to quickly get out of the chaos situation. When the democratic leader knows the team well, such incidentally authoritarian leadership is normally well understood and accepted by the team. The "Dialogue" mentioned in Fig. 4.9 also means information sent by e-mail/Intranet both inwards to the leader and to the other team members on the same level.

A leader who is present much of the time in the project work and who moves around in the project speaking with the team members, who performs *Management by Walking Around* (MBWA), has natural possibilities to notice, at an early stage, when things are not going well and when dangerous attitudes start to emerge. She/he can, in such cases, immediately or quickly initiate changes in the work and gradually change attitudes in the group. Such changes demand pedagogical skills, patience and persistence. The deeper an unwanted attitude has gone, the more time the leader has to spend with the team members to ensure that a sustainable change will take place. Thus, MBWA helps to build a successful company culture.

Thus, by using MBWA (Fig. 4.10), project leaders will feel the real status of the projects long before any written or oral reports reaches them.

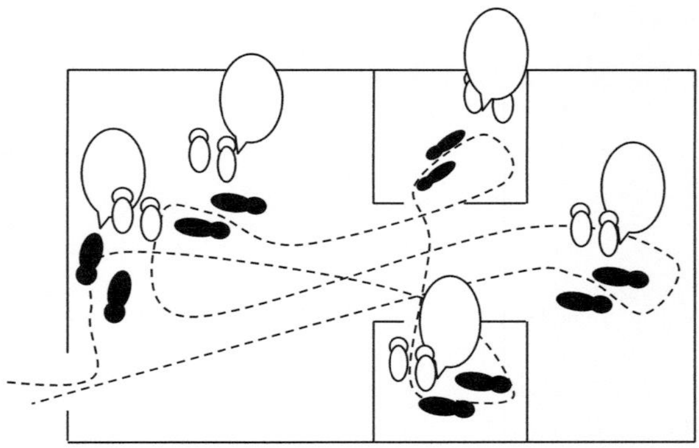

Fig. 4.10 Exercising dynamic leadership means also moving around in the localities and talking with the team members

The project leader must strive to be a few steps ahead of the work, which requires the project leaders and especially the entrepreneurs to also be mobile outside the project localities. This is necessary in order to maintain frequent contact with the customers, society, researchers, experts and other important external people without losing contact with the team members. Thus, dynamic project leaders move back and forth between being in the middle of the project work and being in the "market". They will be proactive more than reactive. They are not afraid of losing control, as their horizontal and vertical feedback systems will provide them with correction possibilities when needed.

A metaphor of how the more bureaucratic/administrative project leader acts and how the dynamic/entrepreneurial project leader acts was shown in Fig. 2.2. In classical mechanics, the same amount of work is needed to push or pull an object, but in reality, there is a big difference in efficiency when the object is a project team. To ensure that the distance does not grow too great between the project leader and her/his sub-project leaders and their teams, it is important not only to have an efficient administrative system and an early warning system when things start to go wrong, but also to have a dynamic control system in which the team members can continuously update what they have done and what they plan to do in the near future, such as one week ahead.

While modern Intranet systems make continuous tracking possible, "writing on the wall", which we will come back to, has shown to be a good complement as not everybody reads messages on the Internet or Intranet and important information, not clearly flagged as such, can be dumped as spam information.

Sometimes, so called *transformative leaders* (entrepreneurial leaders) take care of progress and change while the *transactional leaders* (the managers/bosses) take care of the administration of the existent situation. "Transformative leaders can draw on the interrelated, interconnected tetrad of theory-action-reflection-practice. They recognize that they are, willy-nilly, informed by theory, by ideas, beliefs, and assumptions, and that many of these theories are implicit, meaning they are taken for granted and assumed to be just the way the world is or the way things are always done." (Montuori and Donnelly 2018)

Project leaders, in general, and entrepreneurs, in particular, need to be transformative, as the projects must be transformative to be successful. However, it is not uncommon that, especially in large companies and public organizations (e.g., hospitals), administrative managers are appointed instead of entrepreneurial managers as project leaders.

For the leader to be able to practice a dynamic leadership, she/he should:

- communicate a vision of the goal with the project,
- engage suitable sub-project leaders and let them engage their people,
- draw up strategies and targets as well as giving guidelines for the work,
- break down the work and distribute work activities to groups and individuals,
- ensure that the team members work toward close and distant goals,
- notice and reward good work both on an emotional and practical level.

4.7 Organizational Principles

In larger companies, there is often a staff unit directly under the Managing Director/ Chief Executive Officer (CEO), which is responsible for the portfolio of NPD and innovation projects. At the start of a new innovation project, this unit normally appoints a Product Committee or a Steering Board with responsibility for the organization and financial control of the projects and the portfolio. One of the first acts of such a committee/board when a new innovation project starts should be to appoint a project leader.

When a venture capital firm and/or private investors (often called *business angels*) decide to invest in a new innovation project, they usually also appoint a Product Committee or a Steering Board. We will come back to the financing of innovation projects, but in short *Venture Capital* (*VC*) is financial capital provided to early-stage, high-potential, and high risk startup companies.

In small and medium sized enterprises/companies (SMEs) and in privately started innovation projects, the entrepreneurial project leader normally takes personal responsibility for the development of the innovation concept without appointing a Product Committee or Board. Eventually she/he, will build up a reference/consulting group over time. When the innovation project matures, this reference group can be transformed into a more formal board, which can have credibility advantages, for example, when applying for bank loans.

The steering committee/board decides on goals and milestones. For traditional product development projects following classical rules, timely 'gates' are also set for the project. When gates are used, judgments are made at the gates as to whether a project should get permission to continue to the next development stage or if it should be stopped. At the gates, the committee/board can also decide on reworking one stage or putting the whole project on hold for some time. The number of gates is often not more than five to seven, even for large projects, for example, projects in the automotive industry.

In dynamic product and innovation development projects, the outsider steering committee/board is exchanged for an insider steering committee/board that takes part in the development more or less on a daily basis instead of only at gates. In this case, deviations from plans can be known early as a result of the close contacts with the daily project work, which means a less risky project approach. As the steering committee/board also has co-responsibility for the outcome, instead of just judging the results at the gates, the risk of project failures will decrease. Quite simply, the board directors should be proactive instead of reactive as is the case with the classical board directors.

To secure direct and undiluted information between the Board and the project leader/entrepreneur, she/he should be a full member of the board.

Ideally one board member with experience from the implementation and commercialization of new products and one board member with experience of new product development should act as "*Comets*" in the project. Other life-experienced comets can also be used to secure that there is a strong coupling between the board and the project work.

The Comets move freely around in the innovation project, acting as mentors (see Fig. 4.11), with the job of supporting the project leader and ensuring that the board members have the right information to make critical decisions. Another important mission for the comets is to give suggestions built on their experiences and to improve, in a narrative way, the knowledge of the team members. In addition, the comets will help find out when "enough" good levels have been reached in the development, which otherwise can be a problem for inexperienced team members and team members with low self-confidence.

The interaction between the board and the project in the left hand conventional innovation project organization in Fig. 4.11 is limited to the meetings at the gates, as shown in Fig. 4.12. The example shown is based on the presentation of actions in need-based development.

In general, the conventional PD methods do not discuss sales of the innovations at all, or else they regard sales as something that comes late in the development chain (Holmdahl 2007). However, early sales are of utmost importance for an innovation project as they provide faster and safer development than if a real customer and/or a real user is not directly or indirectly involved in the development.

4.8 Project Localities

When we enter into a room, we immediately form some relation with it depending on what we see of the disposition of the area, the furniture and other objects in it (Nordin-Hultman 1998). The feeling is also affected by how our other senses are influenced by the smell, heat, humidity, or sound. In addition, the total feeling is influenced by who else is in the room and the physical and mental distance to them. Thus, it is a very complex interplay that exists between the room, the things (the artifacts) in the room, and the individuals in it. As a further complication, the total

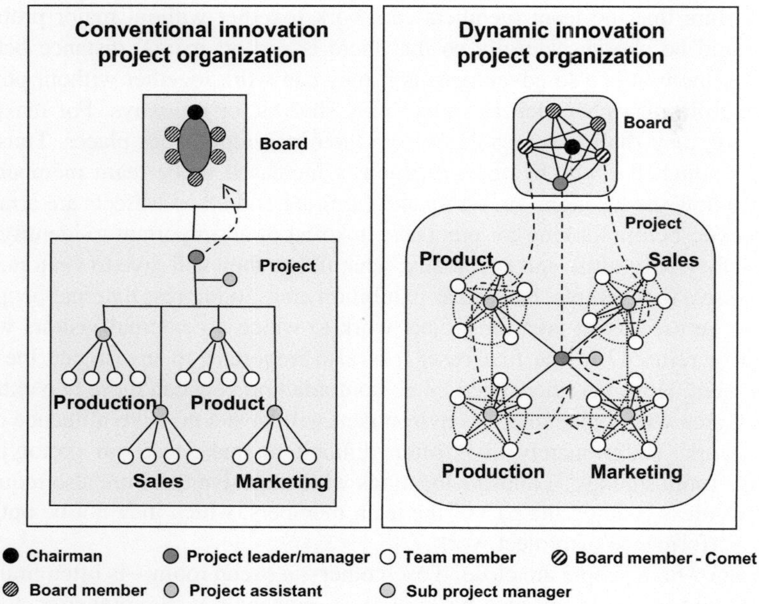

Fig. 4.11 In the dynamic organization, the New Operation/Business Board is an insider board of the development project, which is the opposite to that of a conventional organization

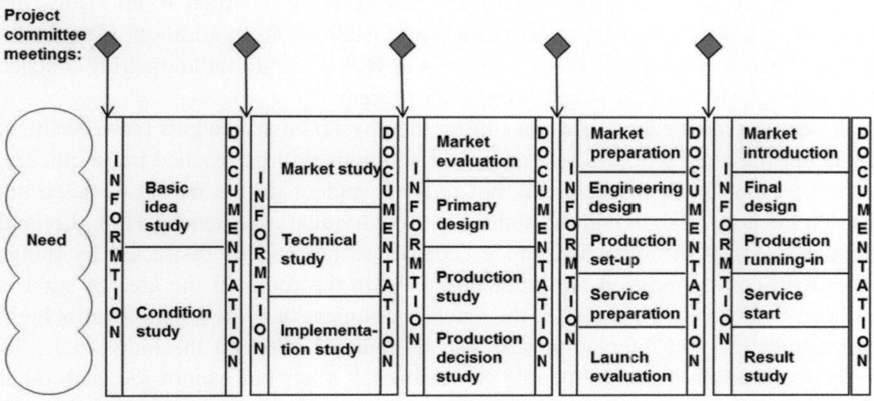

Fig. 4.12 The interaction between the Project Committee and the project for need-based development limited to the gates (the small black rectangles) (Olsson 1976)

feeling is time dependent. For example, the mood will vary for the individuals in the room. The special feeling towards a room deepens or matures when we move around in it. After some time, we get used to it and do not reflect much on it.

To ensure that the team members can work together without major problems, they should be chosen carefully so that there is a short mental distance between them. For them, it is also advantageous if they can work together without physical distance from obstacles such as walls, book shelves, or stairways. For maximum well-being, they shall also be able to form their individual work places. Thus, as a general finding, the well-being in a group is increased if the team members can influence how the locations are set up and furniture and other artifacts are arranged.

However, before looking for premises, it is first of all important to identify what function they will fulfill, and eventually, what image they will give to visitors. Shall the premises, for example, be used as exhibition areas to impress external people, or will they be used as an area for project work to which no external visitors will be allowed or refused? In the first case, it is also important to investigate the outer environment. However, there need be no contradiction between these two extremes as a nice area with a nice interior environment will have a positive influence on the project work. Unfortunately, it is often difficult to make tests on prototypes in exclusive room settings. Exclusive locations with exclusive furniture also require an exclusive dress code on the part of the team members, which may not be optimum for the performance of project work.

A room where people are close to each other—a social room—is often limited in size by fixed walls. It can also be limited by moveable walls, draperies and furniture. Other space-limiting effects are sound, light and climate zones. A light cone and in a big hall lighting up an area will, for example, be perceived as a limited room within the larger space. For example, the lines in a sports arena are regarded by the players as walls around the play area, forming a virtual room. Thus, the mental room does not need to be limited by visual signs. In addition, if someone belongs to a special social group, she/he will feel uncomfortable entering a space where people from other social groups are present.

If people accept each other, the shorter the physical distance gets between them, the better the team work will be and good solutions will be reached more quickly. However, the physical distance is not only dependent on the metric distance but also on the area of sight and the audibility an individual experiences. Humidity and smell also affect the mental distance because people will be distracted by things other than communication and interaction within the room. If the area of sight is reduced by different obstacles, if the smell is not nice, or if the sound level is high, communication and interaction will also be reduced between the individuals. To make work premises where people can hear each other but cannot see each other will cause negative psychological effects on progress.

Different metric zones have been identified, which are dependent on personalities (age, hearing, or sex) and cultures. In the Western culture, we have the following zones (Branzell 1995):

- *Touch zone or intimate zone* (0–1.8 m)—communication through all senses
- *Conversation zone* (1.8–5 m)—two way communication possible
- *Listening zone* (5–11 m)—one way communication possible
- *Mimic zone* (11–25 m)—information through study of body movements

- *Shout zone* (25–70 m)—information through shouting
- *Transformation zone* (70–135 m)—larger movements can be noticed

The more people who are within the conversation zone in which they can see each other without disturbances of sound, light, weather or smell, the better the idea generation and problem solving will be. In the listening zone, "overhearing" will take place when one team member asks someone in the conversation zone for some information. A third person can fill in if she/he has the missing information. Reaching solutions faster can thereby be achieved than if the third person had not been within the listening zone. That is, in fact, an argument that distant work cannot be as efficient as traditional work.

However, old knowledge indicates that often, if more than six persons are within the conversation zone, the interest in taking part in a dialogue will be reduced and the group will split up into dialogue groups with not more than six persons. For example, more guests then six at a dinner table will, for the same reason, split up into smaller conversation groups, which can be an interesting experience to notice when attending a dinner party.

If high book shelves and other big obstacles in a room without inner walls are placed so that the people in it cannot see each other, even although they are not more than about two meters apart, the "we-feeling" decreases radically because the distance, in principle, equals that of the listening zone. Stairways increase the distance as radically as walls do between people working in a normal office with corridors and cabinets. If the doors are closed, the physical distance is equal to infinity—which unfortunately is also the case when the team members are using head-phones to listen to music. If, for some reason, doors must be closed, a somewhat better situation is accomplished for the "we-feeling" if the doors and walls have windows. Thus, the "open door principle" is important for the "we-feeling" in a group.

For early product development projects, the choice of location is especially important as a free flow of creative ideas is important, improving the efficiency/speed of the project.

If it is possible, the models and prototypes should be placed in such a way in the project area that the team members can see them when they enter or are in the room. On the surrounding walls, sketches, drawings, plans and messages are hung. For example, PM Winston Churchill already had "war rooms" with maps on the walls on which his generals could make creative plans.

During the work of a project, it is sometimes important to be able to withdraw from the group to study, reflect and think undisturbed in a silent room. The project leader(s) should also be able to have access to a closed room out of sight of the others, for example, when having personal dialogues with team members, the board chairman, or the project owner.

At meeting and relaxing points, such as coffee corners, there should be white-boards and/or paper to use for sketching. Thereby, good ideas and solutions that suddenly appear in a dialogue can then be noted down before they disappear. Through the use of whiteboards and large paper stands, the creative dialogues will be strengthened, after which the useful ideas can be noted, for example, in personal notebooks.

Creativity seems to be reinforced if the work is done in rooms that are not too well organized or perfectly structured. Chaotic situations, it seems, are the mother of creativity and the most creative researchers, inventors, artists, and product developers often appear to other people to work in a total mess. However, it is noteworthy that many of the biggest companies seem to have been started in simple garages (e.g., Ford, Hewlett and Packard) that were as far as possible from well-organized localities.

In practice, one has few chances to choose project locations, which is why one has to make the best of what is on offer. However, if new efficient premises can create new ways of thinking, they can contribute to a good project outcome, which should be taken into account when new project locations are planned. As so often happens, nature has some solutions to offer and from physics we know that the hexagon is the most stable structure. Compound material often has a beehive cake in a light material between two stronger plates on both sides. Therefore, the shape of a hexagon—as used by the bees—seems to be possible to use also for efficient project locations.

By making a small calculation, a regular hexagon with the length of each wall about three meters should allow a team of six people to always be within the conversation zone if they have their work stations equipped with computer screens and other computer-related equipment placed in the corners of the room.

I noticed visiting AstraZeneca in Bangalore, India that they had some years of positive experiences of this type of arrangement, having four such cells separated with one and a half meter high walls in a bigger square room with normal walls (see Fig. 4.13). There are also two rooms for presentations and discussions in this big square room. The laboratories are close to the square room. The work principle is that the experts work most of the time in their laboratories, coming to the cells to read and write on their computers. When they need to discuss in larger groups, they can use either of the two empty rooms.

4.9 Manning Principles

It is common knowledge that when a development team is to be formed, and especially so for an innovation project as it is more complex than other projects, it is first of all important to find the project leader—the entrepreneur, intrapreneur or renovateur. She/he must have a suitable personality profile for the mission, for when things are running well and when unforeseen problems show up, which they tend to do often early in development projects, the personality profile will be critical for the outcome. The view of life and the competence (knowledge, experience and attitude) are other important factors to take into consideration when choosing a project leader and especially an innovation project leader.

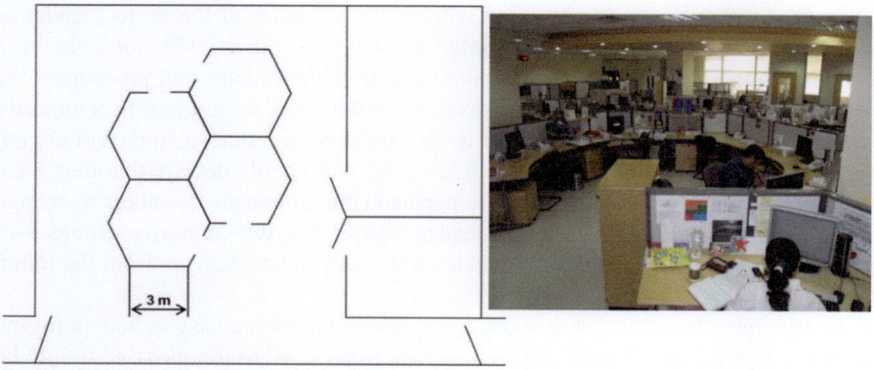

Fig. 4.13 Optimal project rooms are maybe hexagonal rooms with the sides of three meters (photo Stig Ottosson 2007)

It is of extreme importance that the project leader is allowed to choose her/his closest subordinates herself/himself. The same principle should apply when the subordinates start to recruit team members (Lagergren 2003, p. 66). This staffing principle was used, for example, by General Haydon in 1941 to form the mixed troupes that managed to defeat Hitler's troupes in Norway for the first time (Lindegjerdet 2014). Having this recruiting principle means that the personal "chemistry" will work well from the start, as no one is going to choose someone who is difficult to work with or has low work-efficiency records. Thus, it is best not to recruit all team members before the project starts but successively as the project develops.

Unfortunately, it is common that the human relations (HR) staff in larger organizations appoints project members and supplies the project leaders with them. In severe cases, the project leader is even appointed after the project team has started its work. An extremely inefficient start will be the result of such a practice. It will also create a difficult psychological situation when everyone has to find out how everyone else thinks, acts and behaves. Territory will be marked and the level of efficient work will be low for a long time. To reach full efficiency when all team members come simultaneously to the project start can take months, even if the team members are in the same place and have nothing else to do other than the project work.

When a team is to be set up, the different personalities that each team member has will also be of great importance for the outcome of the work. Depending on the unique personality of every individual, she or he will be more or less suitable for the different topics to be dealt with in the project/subproject. Examples of what is required in personal profiles is often that they shall be active, social, flexible, creative, enterprising, positive to change, caretaking, and warm-hearted. However, if the person is not competent in her/his work, it does not help much to have the best soft values.

In a team that has been appointed without the influence of the project leader or when a new project leader is appointed for the team, there will normally be a majority of the project members that are loyal to the principles and procedures the new project leader wants to use. However, there will also, sometimes, be a minority who might say they are loyal but will work with another hidden agenda and not tell the project leader. There will also be a few who will openly declare that they have reached another opinion and will try to persuade the other team members to change direction in conflict with what the leader wants. The two minority groups will appear after some time, especially when reality shows that changes from the initial direction are needed.

In principle, those who openly declare a different view are easy to handle for the project leader as they can either be convinced to be loyal to the ideas or simply be separated from the project. Those who claim to be loyal but in secret work in another direction are difficult to handle and can be seen as "Trojan horses". One efficient tool they will use is to spread doubts and create rumors, which they will ask other team members not to tell to anyone else, creating for them a psychologically difficult double loyalty towards the teller and the team. However, as secrets tend to be interesting to tell, a viral pattern may eventually appear, leading to a negative critical mass that is very difficult for the project leader to change, according to the Chaos Theory (see Appendix A).

For the project leader to be able to recognize Trojan horses at an early stage, it is essential that she/he acts from an insider position in the project as much as possible and to use Management by Walking Around (MBWA) to get a feeling of what is going on. Being in the center of the team and constantly communicating with it gives everyone a better chance to be convinced to share and work in the desired direction. Better ways to work can also be discussed and implemented in an efficient way. By sitting together, overhearing dialogues, discussions, and comments means possibilities for immediate interaction, which, in turn, speeds up this process.

The Trojan Horse is a tale from the Trojan War about the subterfuge that the Greeks used to enter the city of Troy and end the conflict. In the canonical version, after a fruitless ten-year siege, the Greeks constructed a huge wooden horse, and hid a select force of men inside. The Greeks pretended to sail away, and the Trojans pulled the horse into their city as a victory trophy. That night, the Greek force crept out of the horse and opened the gates for the rest of the Greek army, which had sailed back under cover of night. The Greeks entered and destroyed the city of Troy, decisively ending the war (Wikipedia 2013).

Generally, dynamic product development models have a shorter waiting time to get needed information for the team members lacks than classical models. They also give more trust in the team members and let them plan their own work instead of having a central planning department doing the planning (e.g., Ward 2007).

4.10 Project Sizes

Thus, the traditional way of manning a project is to recruit all the people needed at once and have a simultaneous start—a kick off. However, having the people coming simultaneously to a project means that it will take some time for them to get to know each other (to socialize), to bring them all to the same understanding level and for them to learn how the other team members work and which competences they have. Therefore, this way of manning a project can, initially, be inefficient in terms of less well used labor time.

Instead, to get a high speed at minimal costs, as few team members as possible should be used for each activity. In reality, this often means that at the start, the project leader should do all the work her/himself until the workload gets so high that the first team member must be recruited. When more manpower is needed, a team is gradually built up. Then, when less people are needed, the team is reduced so as to always have a minimum of people working in the project. Thus, adding or reducing people when needed and starting every new work sequence with few high performing experts is an efficient principle and is adapted in dynamic innovation management.

As mentioned earlier, most of what has been written about projects has its origin in the study of large or gigantic projects, which is easy to understand as the economic consequences of mistakes in big projects are also big in absolute numbers. Large organizations and large numbers are also fascinating for researchers and management consultants as such organizations represent a lot of power and importance. In addition, there is often more easily available money for studies and improvement in large organizations than is the case for small projects and small organizations.

What especially characterizes large projects is that they take a long time to complete and that many factors will influence the result. Therefore, it is difficult to draw conclusions from them on a detailed level, which is bad as the "devil is in the detail". Also the *Chaos Theories* (see Appendix A) tell us that small changes are the origin of early exponential development until a critical mass has been reached, at which point it is difficult or impossible to calm the situation down.

However, findings from large and gigantic projects are that sub-projects of no more than six to ten team members are the most efficient (e.g., Quinn 1985; Cusumano 1997; Tabrizi and Walleight 1997). Therefore, we should keep in mind that efficiency seems to decrease rapidly when more members are added above six to ten team members for each sub-project. This is especially true as communication becomes difficult when the number of communication routes increases with the addition of new team members. An exact mathematical formula for the number of communication routs "C" depending on the number of team members, "N" is shown in Fig. 4.14 together with some examples of the effects of more people added to a project.

Thus, the number of communication routes grows by the square of the number of team members. According to this formula and from practical experience, it is important to break down projects and sub-projects with many team members into

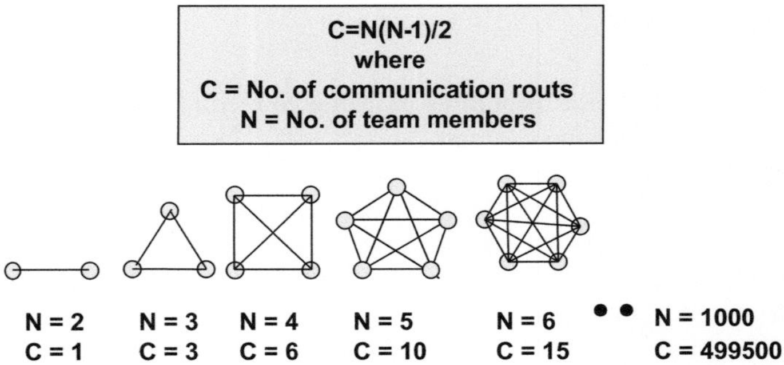

Fig. 4.14 The number of communication routes is dependent on the number of team members

smaller sub-projects and give each group specialized tasks to do. The ISO 10006 norm for projects uses the term WBS—Work Breakdown Structure to describe the goal of such a dividing process. The project leader and the sub-project leaders, respectively, must be responsible for the break-down of the work in order to get maximum efficiency in each work task.

It is clear that the efficiency factor between people can be quite different even though they have the same education and similar experiences. At Microsoft, it is estimated the difference in productivity between the "best" software programmers and average programmers to be a factor ten (Cusumano and Selby 1995).

A consequence of that, in combination with the formula in Fig. 4.14, is that it is better to have few very skilled people working in a team than having many average team members. This is because the larger number of communication routes means a higher risk of some people not being enough informed and that the information activities will take time. Not so productive programmers often have to communicate more to be able to fulfill a complicated or complex task. The more communication needed, the longer it will take to do the work. This conclusion is also in accordance with the so called *Brooks Law,* which says that: "Adding manpower to a late software project makes it later". Brooks law is surely valid, not only for software projects but for all types of projects.

We have experienced that groups with three team members (one leader and two members) can be extremely productive.

4.11 Group Phases

If a simultaneously assembled project team starts its work at the same time, although this is not recommended, it can take a long time to reach full efficiency. This is the case even if they are localized together and have only the one project to work on. One reason is that teams that come together at the same time will have to go through different stages to get to know each other before they can start to work efficiently.

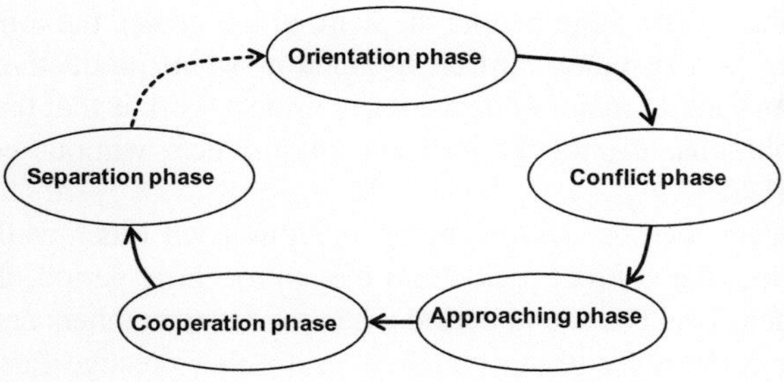

Fig. 4.15 The five phases a team experiences when a simultaneous start is arranged with many people who do not know each other well in advance

Psychologist Bruce Tuckman (1965) first came up with the memorable phrase "forming, storming, norming, and performing" in his article, "Developmental Sequence in Small Groups." He used it to describe the path that most teams follow on their way to high performance. Later, he added a fifth stage, "adjourning" (which is sometimes known as "mourning"). The five phases, re-named in a less catchy and more stringent way, can be called the orientation phase, the conflict phase, the approaching phase, the cooperation phase, and the separation phase (see Fig. 4.15).

During the *orientation phase,* which is a socializing phase, the group members approach each other to get to understand how the other team members work and react, as well as seeking their opinions on different topics. How they finally get together will depend on the personalities and the 'chemistry' that occurs between the individuals. In some cases, two individuals will feel good together from the first meeting. In other cases, they will feel that: "I simply cannot work with that person!" If a formal manager has not been appointed, one or more group members will take the initiative to be chosen by the other group members to be their informal leader.

When the orientation/socializing phase has been going on for some time, the focus is moved to doing the work for which the project has been set up. As the start of a project contains many decisions and as each decision has many options, conflicts will arise when people start to argue for their view of how and in which order things should be done. The team has now come to the *conflict phase*. If the team members cannot settle on the basic decisions and if a strong project leader does not make the decision, sub-groups will develop. In that situation, hostility may arise between the sub-groups, eventually leading to the point where they will not even speak with one another.

The conflict phase can develop into the *approaching phase* when some of the team members take the initiative to solve the problems that have been created between some other team members or sub-groups during the previous phase. Striving for harmony can result in everybody holding back information and views that may cause new fires to occur if the opinion differences are outspoken.

Normally, however, when different opinions occur, the team members will compromise until a majority is in favor of a particular solution. The more team members in favor of a solution, the better for the group harmony in this development phase. The more people there are in the group, the longer it will take each time there is a dispute to reach a situation when productive work can be resumed. A drawback of majority decisions in project work is that they often lead to less radical solutions, eventually resulting in products without edge or unique selling points (USP).

When the team members have gotten to know each other well after the first orientation period, the conflict period and the approaching period, the *cooperation phase* takes place. This is a sound period when the team members can give and take information from the other group members in a rather prestige-free way. It is not until this phase has been reached that the real work can take place. In turn, this means that the three preceding phases have often caused the team time delays so that it can be far behind schedule when they finally reach the situation of being able to work well together.

When the group has completed the task, when it is obvious that a restart with changed goals is needed or when there are evidences that the project will not succeed in reaching its goals of different reasons, it is time to close down the project. The *separation phase* has now been reached. The separation phase can be like divorcing the other team members. For the team members to be able to go further in a healthy way, a lot of energy should be spent on integrating memories, working through separation agony, and planning for a life outside the group. If the separation phase gets too long, the problems will often be greater than if the separation phase is forced to be short.

It is noteworthy that the separation phase does not exist for an innovation project as it transforms into a business process or an operation for which it is important to keep the individuals and their knowledge together for a long time.

4.12 Control Structure

A traditional project can be likened with an English pub walk in the country side, passing fences and hedges with gates and stepping stones where it is wet (see Fig. 4.16). The gates separating each stage are normally judging certain important milestones set in advance. At the gates, decisions are taken if the project shall get a "Go" to continue or a "Kill"/"No go" meaning that the project is cancelled (e.g., Cooper 2001). To control that the gates are not passed before a "Go" is given, larger organizations have sometimes appointed "gate watchers".

From the classical point of view, a project leader often reports to a sponsor or sponsor/steering group/board independent of which type of project it is. The sponsor controls/owns the project specification, or the project contract, or the contract book. Examples of what project specifications contain are (Ulrich and Eppinger 1995):

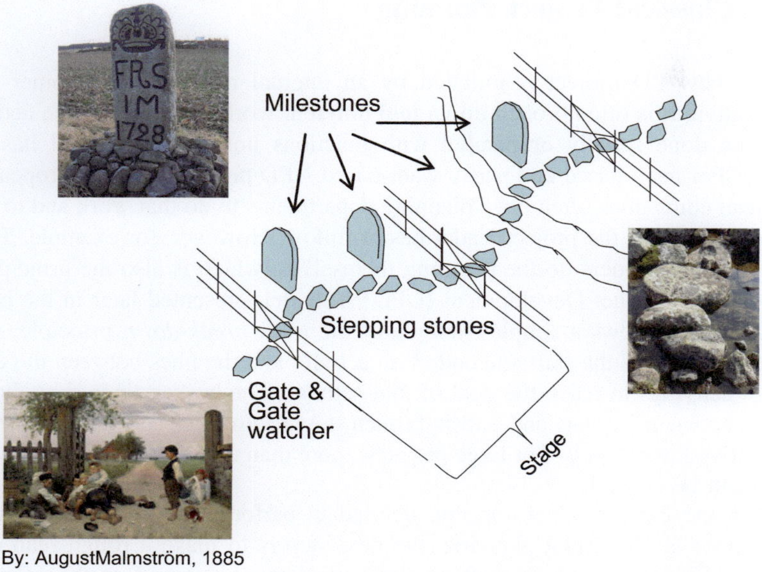

Fig. 4.16 Metaphorical project terms

- A mission statement, the overarching goal and purpose of the project, operational intent, estimated market potential as a basis for prioritizing, etc.
- What specific objectives to achieve
- Action plan (by what actions the objectives are reached)
- Timeline
- Team staffing and project organization
- Resources (budget, etc.), contact persons, partners
- Reporting: to whom, what, when, and how

Sponsor meetings, where the sponsor controls the agenda and sets priorities, decide on project specification changes, as well as controlling eventual additional funding, extends or restricts the project leader's authority, and generally sets the "rules of the game".

However, the control at the gates, when stage-gate principles are used, is too rough in practice, which is why continuous tracking of a development project with regard to P(rice)-C(ost)-T(ime) is recommended. Further, such-real time tracking of the outcome of P and C should be available on Intranet and accessible from any Internet connected computer using a user name and a login code.

When dynamic principles are used, the control is in the hands of the project leader. Thus, she/he has full responsibility for all parts of the project and cannot hide behind board decisions.

4.13 Classical Project Planning

When a new PD project is initiated by an internal or external customer of an organization, it is often broken down into different work activities. These activities are to be done serially or parallel with junctions until the final goal has been reached. For need-based and partly want-based NPD projects, larger European and American companies often have planning departments to do that work and to make detailed plans that the project leader has to follow. However, for example, Toyota lets the project leaders do the planning themselves, which is also the principle for the Dynamic Product Development (DPD™) model, presented later in the book.

Figure 4.17 shows a simple example of the *work break-down* principle, where each circle denotes the start and end of an activity and the lines between the circles denotes activities to reach the goal of that activity. The broken lines show how an activity between its start and end is broken down into finer and finer patterns on consecutive lower levels. For large projects, more than the three levels shown in the Figure can be needed.

When the basic product concept, defined in performance (P) terms, has been agreed upon in the initial planning, the next step is to make a rough plan of the project to find out when it will be finished at the latest and at the maximum accumulated cost. This planned development is often thought to follow an "S-curve" (see the dotted line in Fig. 4.18).

The accumulated cost curve in the Figure is also often thought to follow an S-curve. However, in real development projects there are many ups and downs in the P-curve, while the accumulated cost-curve goes up and rests when no work is done during the life of the project. Often or even in general, the performance target

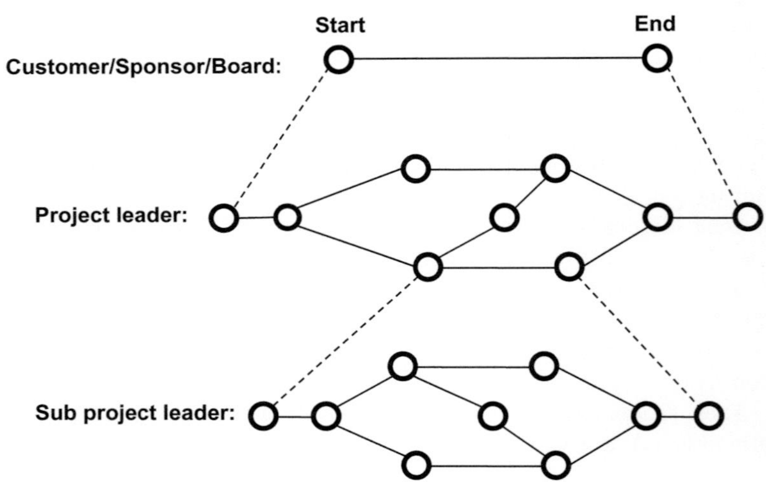

Fig. 4.17 A project normally has three decision levels

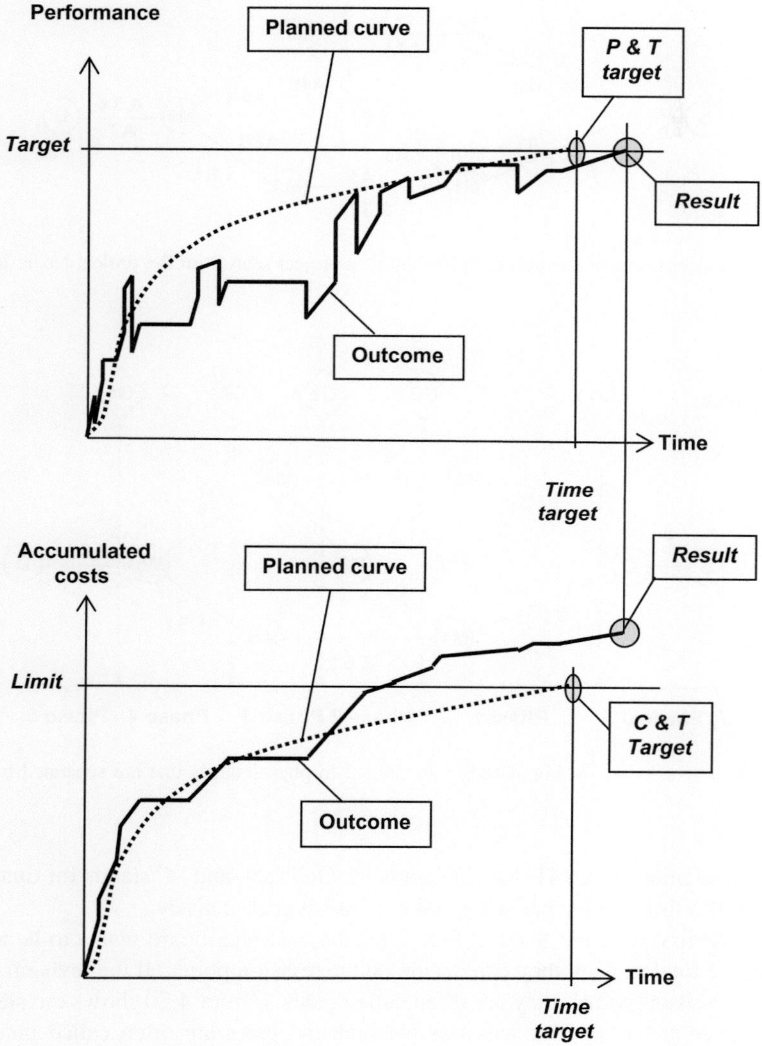

Fig. 4.18 An example of a planned PCT outcome and real outcome of a project

is reached later than was planned. This is because unplanned and unforeseen disturbances occur during the development work.

When performing PD projects, some activities must be done before others are started, while some others can be done in parallel. The goal of an activity can be called a "Milestone". Figure 4.19 shows the situation on the level of a project leader. Here S stands for Start, C stands for Concept, A0–A7 stand for the activities

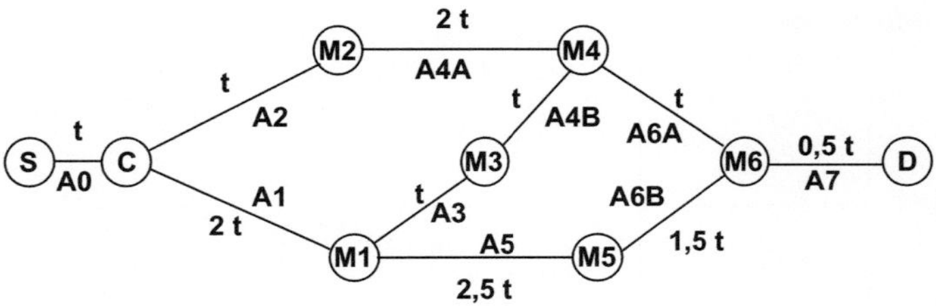

Fig. 4.19 An example of the detailed planning of a project seen from the project leader level in Fig. 4.17

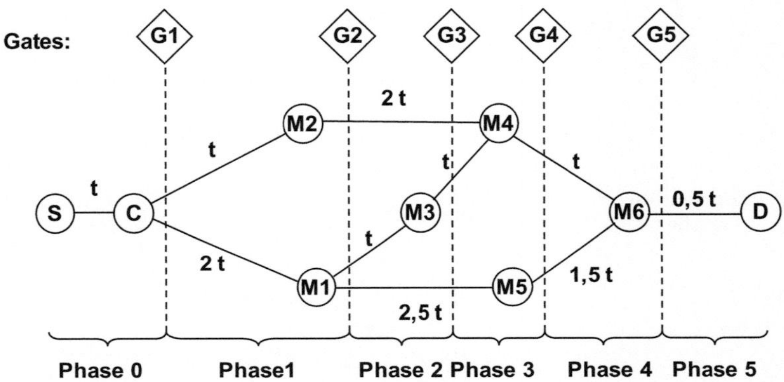

Fig. 4.20 The activities in Fig. 4.18 are assembled to phases/stages that are separated by gates (G1–G5)

to reach the milestones M1–M6, D stands for Delivery, and "t" stands for time units equal to the time it is supposed to take to finish each activity.

If at the first decision level in Fig. 4.18, the customer/board wants to be able to control the total project, they can decide on the revision points. If the revision points are also decision points, they are often called gates. Figure 4.20 shows the situation if five gates are used. The activities between the gates are often called phases or stages.

The gates are normally set after certain important milestones are planned to have been reached. Thus, the gates, which sometimes are drawn as traffic lights, are "timely gates" to control that certain results are reached at a time decided on in advance. The milestones are performance-based. Figure 4.21 shows the principle differences between the terms most often used.

For the planning of each milestone, so called Gantt schemes are often used. *Gantt schemes* are time bars for each activity, giving a picture of how long each activity will take. Figure 4.22 shows the situation for the example shown in

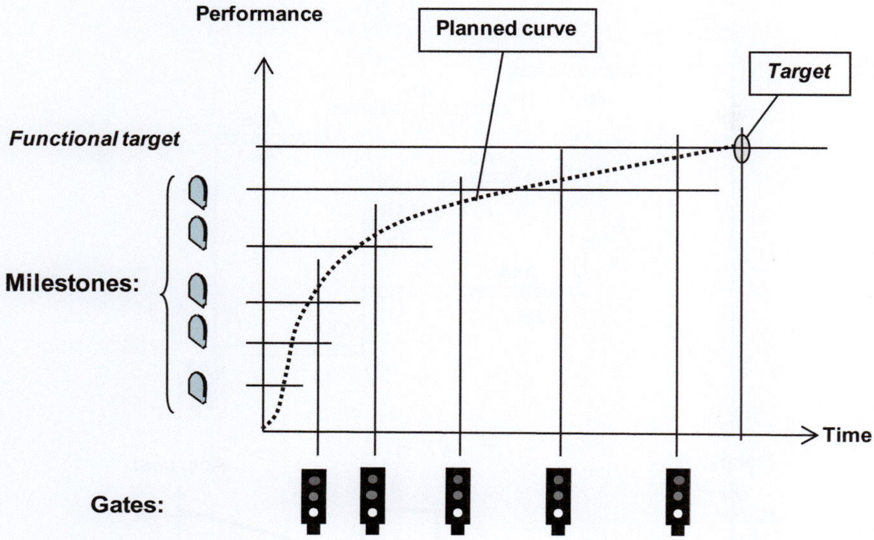

Fig. 4.21 Milestones are connected to performance, while gates are used to control that the project does not exceed the time limits which have been set in advance

Fig. 4.21. In the lower picture, the number of people needed in order to finish each activity is shown. The accumulated cost based on labor costs is also shown in the lower Figure. Note that, unfortunately, gates are difficult to show in Gantt schemes.

Another way to present the situation shown in Fig. 4.20 is represented in Fig. 4.23. This method is called the Critical Path Method (CPM). In the CPM presentation method, the Northern sectors of each circle tell which milestones the circles represent. The western sectors tell the accumulated time from start for the longest time of two or more branches that join at the circle. In a similar way, the eastern sectors tell the calculation backwards from the delivery time. The southern sectors tell the waiting time, which should be the same as when the Gantt scheme is used. The critical path is the path that takes the longest time from the start to a finished project. If one highest (optimistic—o) and one shortest (pessimistic—p) time is given for each activity in addition to the most probable time (m), the foundation is given for PERT—Program Evaluation and Review Technique. The time for each activity is calculated as: $t = (o + 4\,m + p)/6$.

The best situation for a project is if there is no waiting time at each junction. However, to shorten the time for activities in the critical path means added costs, which is shown in Fig. 4.24.

The corresponding CPM scheme is seen in Fig. 4.25.

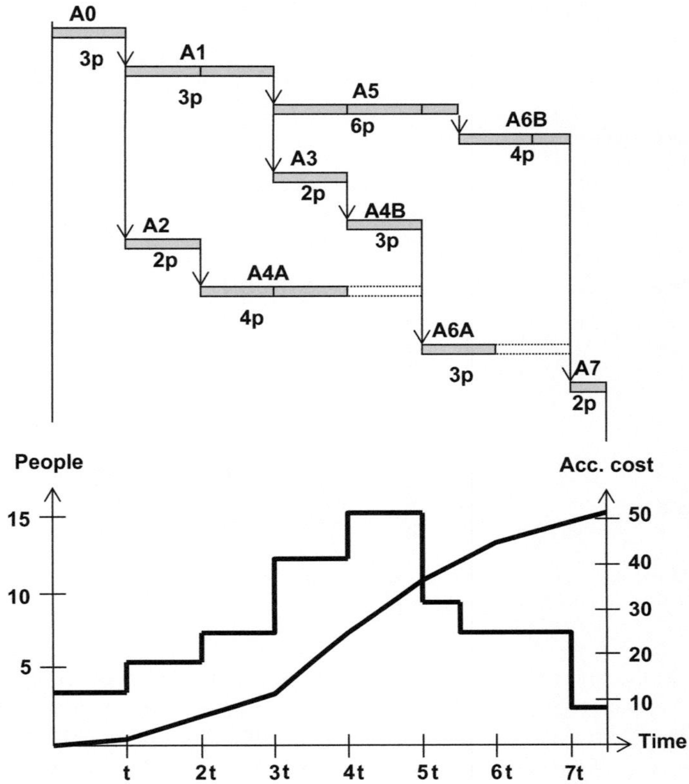

Fig. 4.22 Figure 4.21 represented as Gantt scheme with added accumulated cost estimation (p stands for persons)

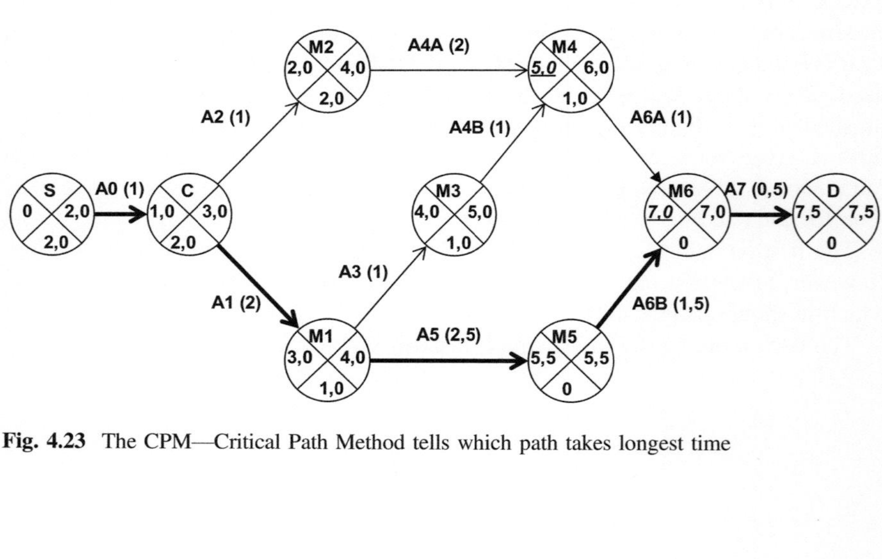

Fig. 4.23 The CPM—Critical Path Method tells which path takes longest time

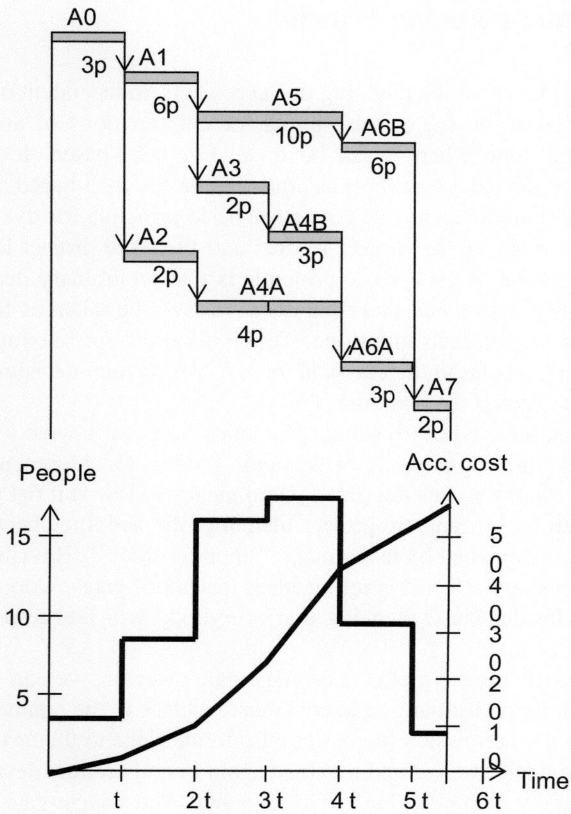

Fig. 4.24 The figure is trimmed to get rid of waiting times causing a shorter total project time on account of higher accumulated costs

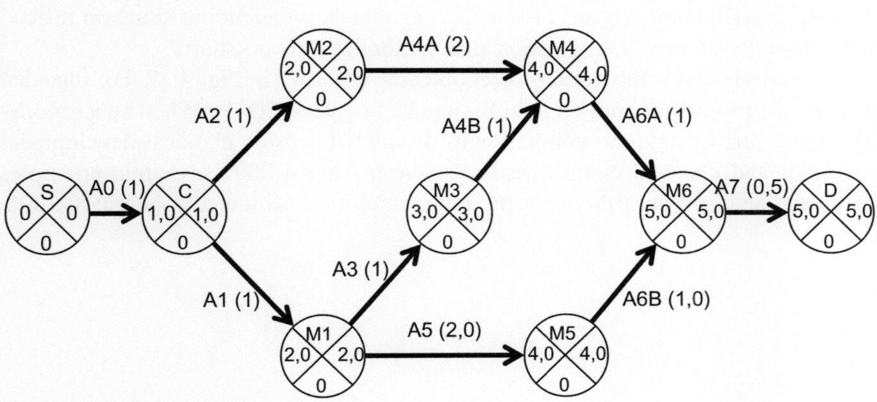

Fig. 4.25 The corresponding CPM scheme to Fig. 4.24

4.14 Dynamic Planning Principles

A basic principle for dynamic planning of PD projects, independent of whether they are need-based, want-based, or wish-based, or combinations of all three, is that planning will be done when it can be done. For need-based development, the development horizon can be as short as hours, while for wish-based situations, it is, in general, long (months and years). Another basic principle for dynamic planning is that it will be done by the project leaders and their sub-project leaders and not central planning units. A third basic principle is that preliminary decisions will be taken as early as possible and that final decisions will be taken as late as possible before the dead line of each milestone. This is to allow for maximum flexibility. This principle is opposite to the classical view in which final decisions are taken as early as possible before the deadline.

Gates are not used when dynamic principles such as Toyota's Lean Product Development, Agile Programming or Dynamic Product Development (DPD™) are used. This is because the outsider gate system means a slow and risky development situation. Therefore, in dynamic project planning, the activities between the milestones can be accomplished by using many "stepping stones". Revisions are used to check on development when it is felt needed. Instead of gates, workshops are used to increase the insider view, meaning that everybody will feel responsible for the development.

To underline the negative effects of stage-gate systems, we can see that if few gates are used, both performance and cost must be close to the planned curves if the project is to get the permission to continue from one stage to the next. However, as milestones are gained and lost and regained again in real product development, it is, in reality, extremely difficult to get a "Go" at a gate. Thus, stage-gate methods mean that a "Kill" or "No-Go" will be the most probable decision at each gate except for at the first gate. This is because a rejection is stronger than an approval (if there is a "Yes" for the performance and a "No" for costs, the combined decision will normally be a "No"). Figure 4.26 shows the difficult situation when few gates are used. The outsider situation at gates also creates an attacking/defending situation instead of the friendly cooperation situation that symbolizes a workshop.

An example of a want-based project outcome is shown in Fig. 4.27. For financial reasons, the project was put on hold for almost one year (2004). When new funding was found, the money was coupled with demands for using classical development principles and classical control demands. From August 2007, dynamic principles were used again to get to a ready product, which was achieved in February 2008.

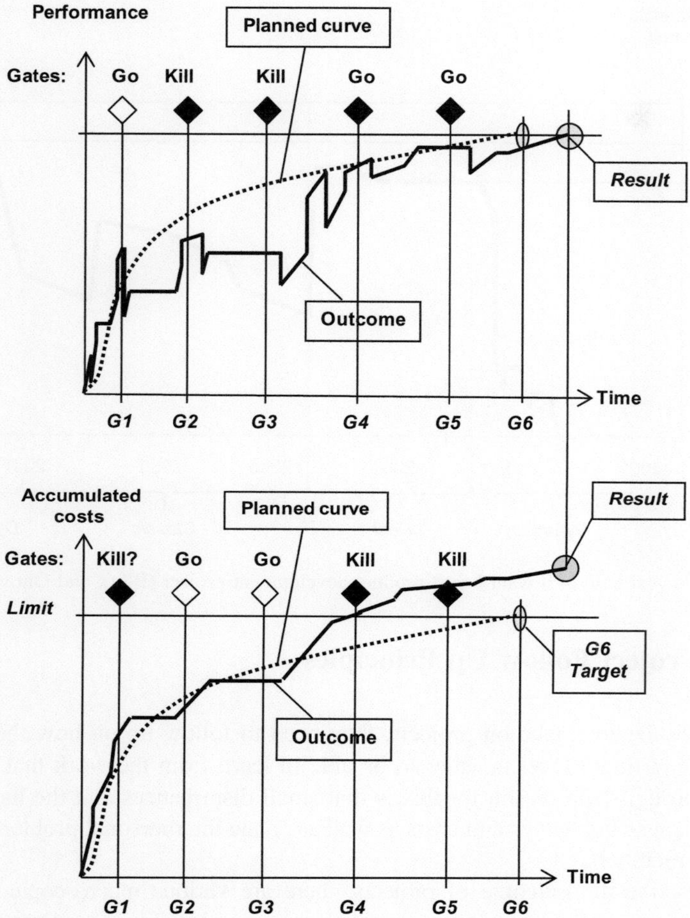

Fig. 4.26 If few gates are used, the probability will be high that a development project will be stopped—"Killed"—before the targets are reached

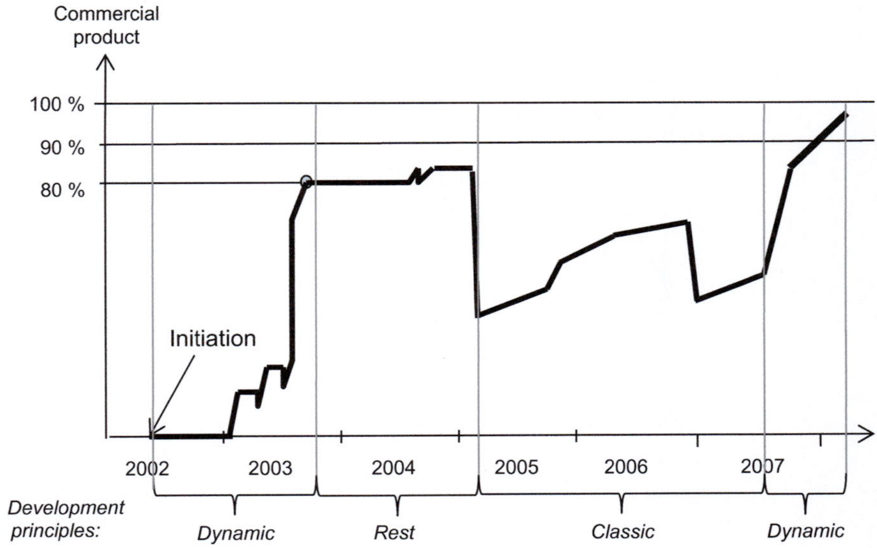

Fig. 4.27 A real case of a want-based product development project (Björk and Ottosson 2008)

4.15 Project Follow Up Principles

When organizations take on projects, they need to follow up on how the projects worked on a micro level in order to be able to learn from the work that has been carried through. The reason for this is that small disturbances and the handling of them can cause big delays and costs as well as create interpersonal problems among the team members.

To describe the outcome of projects, there are various macroscopic graphical follow up systems available on the market. Some examples of such systems are shown in Figs. 4.28–4.30. The background of these and similar graphical presentations is that

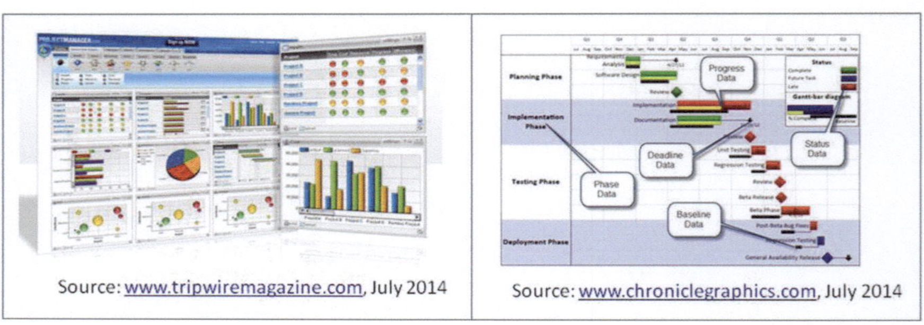

Fig. 4.28 Examples of commercial follow up representations of projects

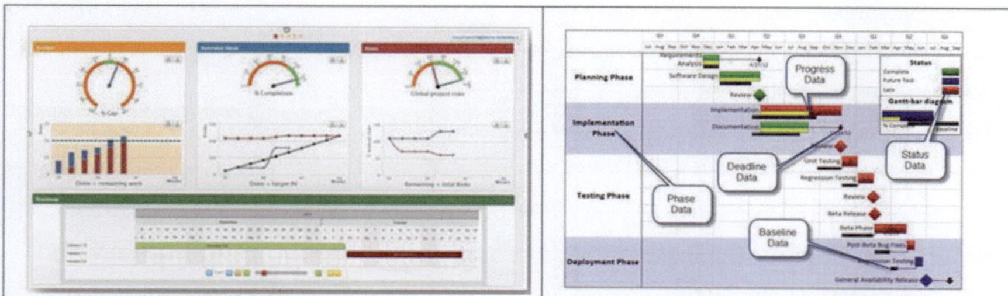

Source: www.qualitydigest.com, July 2014

Source: www.chroniclegraphics.com, July 2014

Fig. 4.29 Examples of commercial follow up representations of projects

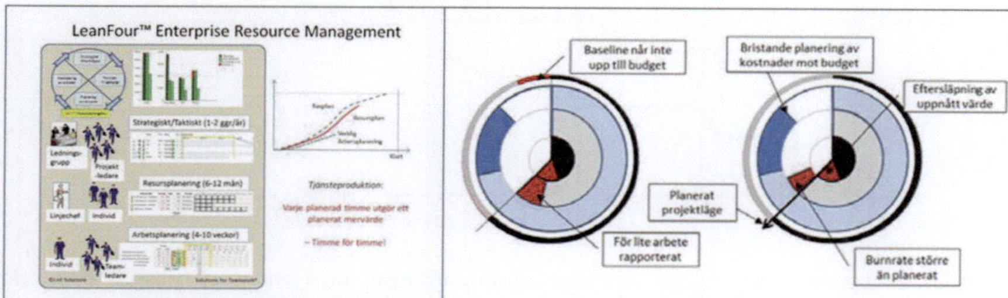

Source: www.ln4solutions.com, July 2014

Source: www.ln4solutions.com, July 2014

Fig. 4.30 Examples of commercial follow up representations of a project

they seem to build on the idealistic opinion that projects follow the S-shaped curve shown in Fig. 4.21, which is seldom the case in reality.

Unfortunately, macroscopic graphical follow up systems give little information on why a project was not successful. Therefore, to find out why a project, that was supposed to last for two days took much longer time, a thorough investigation was done afterwards on who did what in the project. The representation is shown in Fig. 4.31.

From Fig. 4.31 the following conclusions can be made:

- The project leader (TS) did not lead the project
- The deputy PL HH did not coordinate the activities
- SO became the informal leader

Because of this, the external costs became three times higher than needed and the time it took to finalize the project was also three times longer than needed.

This small project showed that the graphical method used in Fig. 3-32 can be used for evaluating what has been going on in a project and can also show when

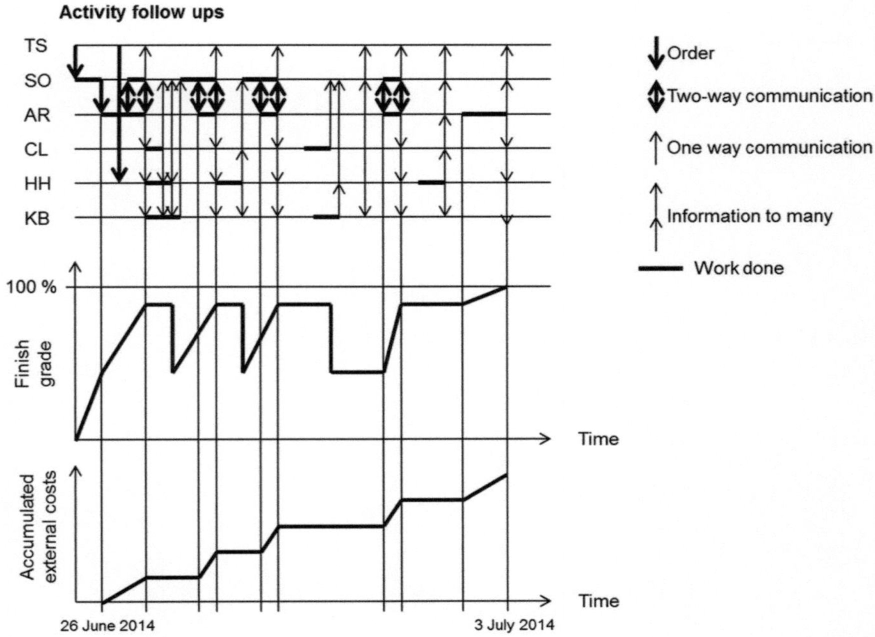

Fig. 4.31 A representation of the communication and work in a one-week project (Ottosson 2014)

actual work was done and how communication took place. However, there are some limitations with the method when a manual follow up is done, such as:

- it will be difficult to use for many participants—probably ten is the practical limit
- evaluation should be done in short time intervals (1–2 weeks)
- it works best for distributed teams for which communication is done via IT.

Chapter 5
Knowledge Generation

5.1 Knowledge Generation

For every organization, both individual and collective knowledge are critical resources.

All around us, there are different types of data (see Fig. 5.1). For data that catches the interest of someone, this data becomes information. Further, if one reflects on the new information it becomes knowledge. If one critically revises the information, one will be wiser than if one "buys" the information right off.

Note that what is common for all information is that it can—and is—manipulated to be what someone, whether for good or bad reasons—wants it to be. This also counts for scientific papers and articles. Reality is constructed from our thoughts of reality and there is no reality until that reality is perceived. Furthermore, no clear dividing line exists between ourselves and the reality we observe to exist outside of ourselves (Wolf 1989, p. 128). Instead, reality depends upon our choices of what and how we choose to observe. These choices, in turn, depend upon our minds or, more specifically, the content of our thoughts and our mood, intentions, our interest, our expectations, and our desire for continuity. Also, not to be forgotten is that the more we determine one side of reality, the less the other side is shown to us!

When comparing and combining the knowledge with earlier gained knowledge, our own understanding will grow. Then, if one uses the understanding for action, the outcome of the action will be new data.

Again, some of the response data will be of interest for reflection and new actions in loop after loop and will increase knowledge as practical wisdom. Theoretical education and practical training will help an individual to start at a higher loop level when creating new knowledge. If she/he also has a broad range of interest outside the actual field, this will help to make the person more cultivated—which is a great asset, especially for an entrepreneurial project leader.

S. Ottosson, *Developing and Managing Innovation in a Fast Changing and Complex World*, https://doi.org/10.1007/978-3-319-94045-8_5

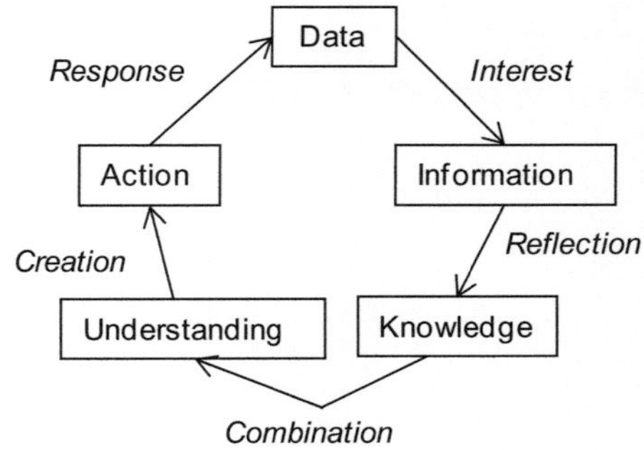

Fig. 5.1 The knowledge generation loop

To deepen our understanding of something new, we take different actions, such as manipulating an object. The response will be that the object emits—or signals—data. We can also manipulate the environment, in which the interest is to see what kind of reactions we get from our different manipulative actions. To help us notice and evaluate, for example, weak signals, we can utilize "machines" that have been programmed to give us wanted information. We often prefer to call such information "artificial", implying that machines can think in a similar way as human beings do, which so far is not possible although the development of artificial intelligence (AI) is rapid. However, a thought is not only a phenomenon in our brains but also a result of how our brains actively interpret our experiences in confrontations with the surrounding world.

5.2 The Learning Process

All innovation and New Product Development (NPD) activities are both individual as well as collective learning processes.

Thus, for everybody taking part in a development project, a personal experience and knowledge gain will be the result every time the person thinks or works on the project, given that the person critically reflects on her/his experience. If the experiences are discussed with other persons in the team, the *collective knowledge* will increase. Knowledge is gained both when things go well and when things do not go so well—which is often the case, especially when a radically new product is developed. The knowledge is accumulated stepwise.

To help remember personal knowledge gained, it is good to make notes in your own notebook from meetings when ideas or problems have occurred. Looking back in the notebook now and then gives possibilities to reflect on the work done and why certain decisions were taken. The personal wording in the notebook is also probably more useful for such reflections than notes written down by a meeting secretary. Therefore, it is recommended to make personal notes, also from formal meetings.

Paper notebooks have some advantages over electronic notebooks as, in general, electronic technologies are not specifically designed for engineering and, as such, do not provide comprehensive support for the full range of essential engineering tasks, such as freeform sketching and annotating external documents (e.g., CAD drawings and PDM files). Other more routine but important practical requirements, such as portability, survivability and start-up time may also prove to be fundamental barriers to the use of an electronic logbook. Consequently, the use of paper notebooks is still recommended.

Notebooks also have a legal implication if patentable inventions have been produced in the work. If many apply for the same solution at the same time, the notebook can be the evidence needed to save the patent rights. Therefore, it is important to also note the dates and to use a ball pen rather than a pencil when making the notes. This is because information can easily be changed afterwards if pencils have been used.

Even when electronic storing is used, paper and physical copies should be saved in case the digital media breaks down or becomes inaccessible in the future, for example when the technology changes. Thus, tape recordings from the 1980s are difficult to access today and diskettes from early 2000 cannot be accessed with today's PCs. Also CDs, DVDs, and Blue-ray Discs (also known as Blu-Ray Discs) are difficult to access or are impossible to access on new Laptops and PCs.

Note that a legal strong situation only exists if electronic data cannot be changed afterwards, which makes most computer programs not suitable for such storage. Further, if accidents happen with the products while in use, the notebooks can be valuable source of information in court trials to find out whether the fault was caused by misuse or product deficiencies. Going back to the notebooks can give evidence as to why a certain solution was chosen.

To be valuable afterwards, notebooks must constantly be updated as close as possible in time to the findings. A big problem, however, with making notes is that it demands a great deal of discipline. However, with the possibilities offered by modern digital voice recorders, information can be transferred to one's computer for automatic translation to written text. This makes it easy to make notes, although sketches cannot be done in this way.

For some development companies, such as for the drug industry, logbooks are requested by the authorities. Such logbooks are normally written by a project staff member or a project assistant for smaller teams of R&D personnel. The logbooks are normally a summary of what a team has accomplished, summarized in diagrams and sketches. Support for the work is provided by the notebooks, together with dialogues with the individuals who have written their findings in them.

5.3 Personality Aspects

Everybody has her/his own competence, experience, personality and "ausstrahlung" (i.e., charisma, image, status, etc.). These are interconnected in a complex manner. They can also be divided into many sub-parts. Competence, experience and personality are mostly hidden until they show up in the action of one kind or another.

The ausstrahlung is weakened or strengthened by the actions one makes. The ausstrahlung a person has is an asset, for example, for getting other people to cooperate better in a development project. It can also make people less reflecting and uncritical, even to the point of blindly following, for example, sect leaders in a way that is difficult to understand. Unfortunately, education gives no protection against this negative side of ausstrahlung from formal and informal leaders (McMahon et al. 2002).

Personality is not constant over time, but will change depending on the situation and the people an individual comes into contact with. In an un-pressured situation, someone may have one type of behavior. Under stress or threatened, that same person may have another. Figure 5.2 shows an example of how the experienced personality for one semi-professional soccer player in a field test changed depending on the situation. The dotted line shows his behavior when he was physically and mentally exhausted. (IQ in the Figure stands for Intelligence Quotient, EQ for Emotional Quotient—social competence—and AQ for Adversity Quotient—survival competence). When recruiting managers, EQ-tests have become more and more common in selecting between the applicants for a profession.

The two lines in the Figure are based on how the coach of a soccer team viewed each player in his team before and after a stressful situation. Thus, the solid line shows what the coach regarded before the whole soccer team was engaged in stressful situations for three days and nights, and the dotted line shows what he had experienced following the team during the three test days and nights. He made the same curves for all his players after the tests and could afterwards set up the team better, depending on if they were leading a match or not. The team became successful after having had great difficulties before the test was done.

When a project team is to be set up, different personalities and the ausstrahlung of potential team members have to be discussed as they will be of great importance for the outcome of the project. Also, depending on the unique personality everybody has, she or he will be more or less suitable for the different topics to be dealt with in the project. Examples of what is required in personal profiles is often that the team members should

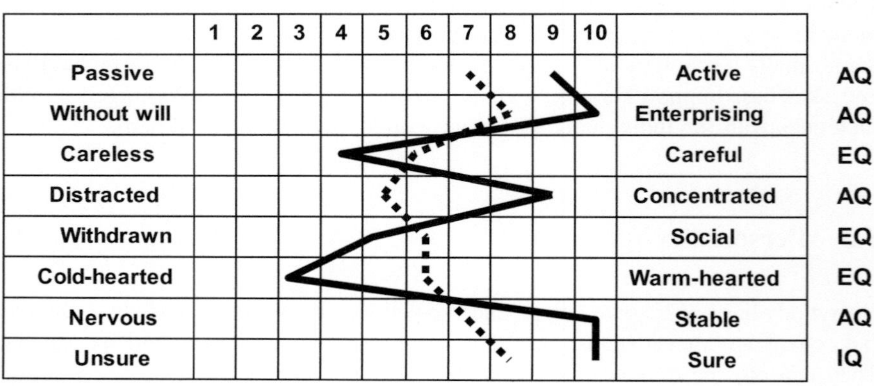

Fig. 5.2 The solid line shows the behavior of a person in relaxed situations

be active, social, flexible, creative, enterprising, positive to changes, caretaking, and warm-hearted. However, if the person is not competent in her/his work, it does not help much to have the best of such soft values. For the evaluation of potential team members and the setting up of the team, a Pugh matrix (see Sect. 7.9) can be used.

Important for setting up a team is also how well two or more people will interact dependent on the task, the personality profiles that each of them have, and how well they will go together—how well the personal "chemistry" will work.

The composition of a team has proven to be of extreme importance and it is said to affect team performance 40 times more than team coaching (Hackman 2002, p. 208). However, there is also a risk of achieving too little cohesion in a group of disparate talents.

If the personality differences are too big, communication within the team can be hampered (Sample 2004). Also, it should be considered that in a homogeneous team there is a tendency for less learning and for low creativity in the team (Holmdahl 2007).

Therefore, it can be concluded that it may be advantageous to recruit team members that complement each other's competence and who have a good "chemistry" between them. The combined competence will be larger for such a group than for a team with much the same competence but with little differences in their views of life. Therefore, a heterogeneous group should perform better than a homogeneous group (Pech 2001).

Understanding the different personalities when composing a team is important as this makes it easier for us to learn how to better deal with colleagues, clients and other people. Understanding gives a better perspective of a person and how they react, for example, when faced with a stressful situation.

Experts have suggested that there are four basic personality types and given them four colors. The four basic colors are red, blue, yellow, and green and some characteristics of the four personality extremes are shown in Fig. 5.3. The flashes denote that, for example, red and green personalities might have some problems getting on well together. There is a lot written about color personality psychology (e.g., www.kent.ac.uk/careers/Choosing/colour-personality.htm). Note, however, that each of us has a mix of colors in ourselves (Erikson 2017).

5.4 Competence Aspects

Competence is often seen as a combination of knowledge, skills and attitude. Competence grows through experience and an individual's ability to learn and adapt. When we increase our knowledge we gradually will become more understanding, which is helpful for improving our skills in developing, for example, new solutions. Thus, knowledge is vital, but without understanding, we will not be able to be productive. Having a negative attitude will be harmful when it comes to enhancing our total competence.

A person's competence will be experienced first when she/he is in action in a situation in the environment in which she/he operates. Such an environment is often called the *context*. In emergency contexts, people will react to the situation following

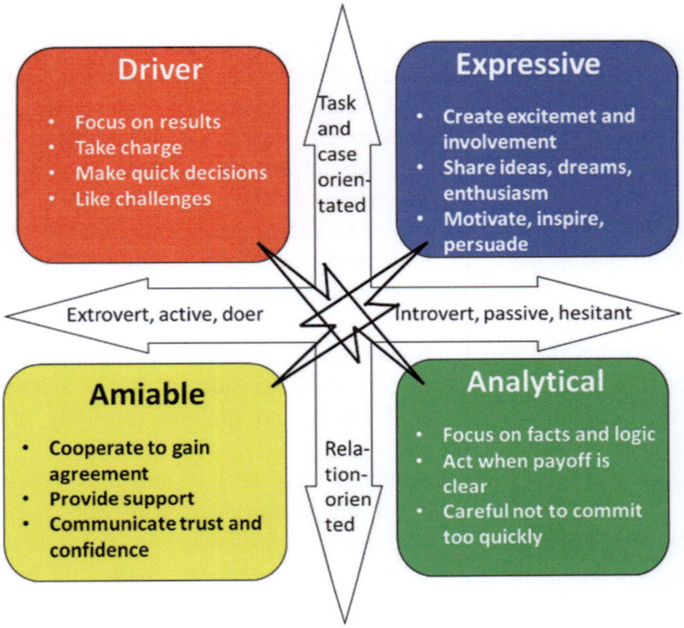

Fig. 5.3 The four extreme personality colors

behaviors they have been trained for and/or previously found to be successful. In a less time pressing situation, they can try new ways of working that can be successfully used later in an emergency context. Testing and training is part of a learning process in which individual and collective competences are built up. The process of development of individual competence is a lifelong learning process.

To be *competent,* one needs to be able to interpret the situation in the context and to have a repertoire of possible actions to take to manage a situation. When that is not enough, creativity will help to provide input for improvising. This means to do something new to handle the situation when the accepted behavior does not work or the situation is different from that which one has been trained for.

On a detailed level, competence is often seen as a standardized requirement for an individual to perform a specific job properly. It encompasses a combination of knowledge, skills and attitude utilized to improve performance. More generally, competence is the state or quality of being adequately or well qualified and having the ability to perform a specific role. As competence covers so many facets of life it, is sometimes broken down:

- *Meaning Competence*: identifying with the purpose of the organization or community and acting from the preferred future in accordance with the values of the organization or community
- *Relation Competence*: creating and nurturing connections to the stakeholders of the primary tasks. (A stakeholder is a person, group, organization, member or system who affects or can be affected by an organization's actions.)

- *Learning Competence*: creating and looking for situations that make it possible to experiment with the set of solutions that make it possible to solve the primary tasks and reflect on the experience
- *Change Competence*: acting in new ways to promote the purpose of the organization or community and make the preferred future come to life
- *Occupational Competence*: concentrate on self-image, values, traits, and motive dispositions (i.e., relatively enduring characteristics of people) that are found to consistently distinguish outstanding from typical performance in a given job or role.

To distinguish the competence between people, the steps shown in Fig. 5.4 are sometimes used, although these maybe not so nice to see when one is young and inexperienced. However, reflecting on the Figure can give inputs for a personal strategy for how to benefit from other people, for example, when setting up a product development or innovation project.

On a more detailed level, the steps in the figure can be explained as (Dreyfus and Dreyfus 1986):

- *Novice*: rule-based behavior, strongly limited and inflexible
- *Experienced beginner*: incorporates aspects of the situation
- *Practitioner*: acting consciously from long term goals and plans
- *Knowledgeable practitioner*: sees the situation as a whole and acts from personal experience
- *Expert*: has an intuitive understanding of the situation and zooms in on the central aspects
- *Virtuoso*: has a higher degree of competence, advances the standards and has an easy and creative way of doing things
- *Maestro*: changes the history in a field, for example, by inventing and introducing radical innovations

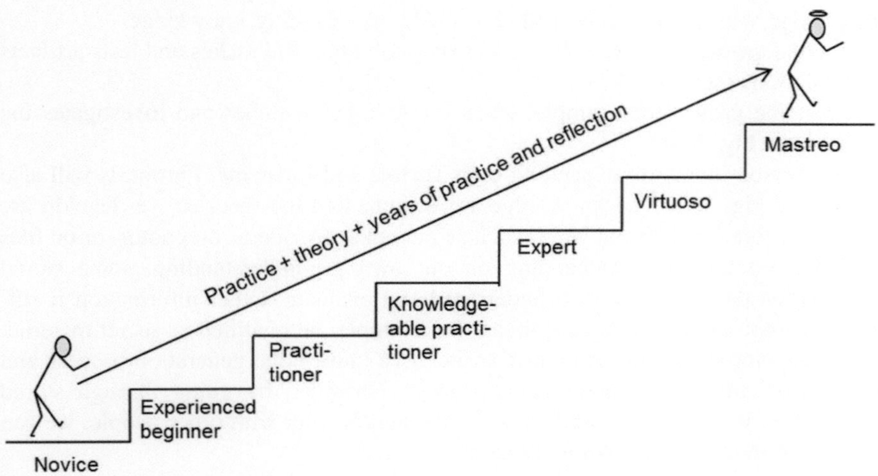

Fig. 5.4 Different steps going from being a novice to being a maestro

5.5 Personal Knowledge

Discussions about the term knowledge have been a philosophical topic for the last twenty-five centuries. Already the ancient Greek philosophers (e.g., Aristotle/Socrates and Platon) classified knowledge into the three classes:

- *Techne*: practical—productive knowledge
- *Episteme*: theoretical—scientific knowledge
- *Phronesis*: knowledge as practical wisdom

Techne is connected to "know how" and Episteme is connected to "know why". While Techne and Episteme grow rather independent of each other, Phronesis grows when both Techne and Episteme grow. Dynamic, receptive and life-experienced people with broad knowledge and well developed intuition are therefore wiser than people who lack one of these characteristics. For this reason, an expert is not also automatically wise.

The personal ability to increase individual knowledge is dependent on many factors, such as one's own personality, intelligence, creativity, ability and feelings. The ability to take in information and to combine it with what one already has accumulated, and thus, to be new know-why and know-how is dependent on the mood, interest, motivation and attitude at each moment. This can be called a filtering effect.

Note that, in this respect, spoken information evaporates fast if one does not make notes or recordings from a speech or a two way communication. Estimations tell that less than 20% of the information that we receive is used to build up new pictures of the world in our brains, while the remaining part comes from pictures stored earlier in our brains (Wheatley and Kellner-Rogers 1999).

Being in a good mood and being interested and motivated means that our own filters are open for us to take in and process most data and information and a maximum of all incoming impressions/signals. Not being in a good mood, being uninterested and being unmotivated means that no or very little information and knowledge will be able to be added to one's own existing knowledge.

Techne grows when an individual creates solutions and makes and tests artifacts and solutions (see Fig. 5.5).

Episteme grows, for example, when a researcher searches and investigates the reality (see Fig. 5.6).

If a person works on expanding both Techne and Episteme, Phronesis will also grow (see Fig. 5.7). The knowledge we acquire and the theories we develop are normally stored in different ways, such as in books, on videos, on photos, or on files available on the Web. Depending on our own pre-understanding, some stored information can be difficult to understand and evaluate. Other information is difficult or impossible to access as it can, for example, be qualified as secret material. Every development process is both an iterative knowledge generation process and an iterative information generation process. However, by going through stored information over and over again and by communicating with other people, we can iteratively improve our own knowledge.

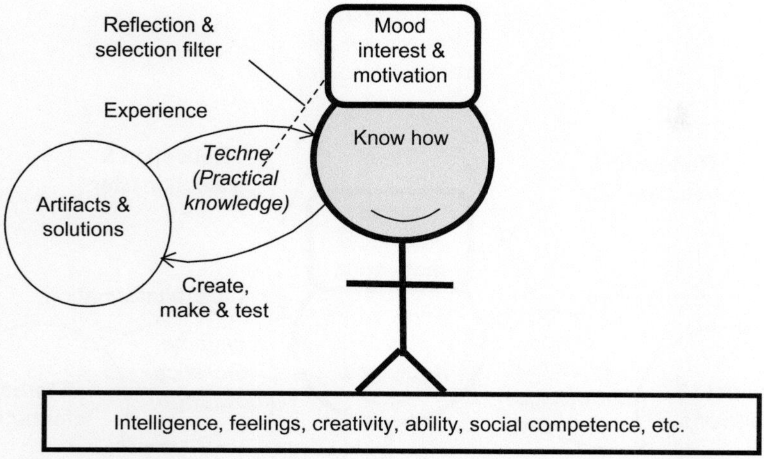

Fig. 5.5 The practitioner's way of expanding techne. Solutions in the figure are, for example models, prototypes, or manufactured products

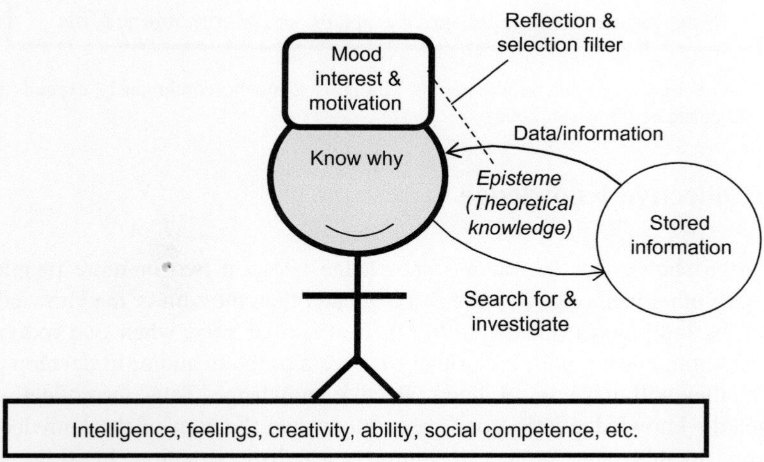

Fig. 5.6 The researcher's way of expanding her/his episteme

As Fig. 5.7 indicates, one important way to extend one's own knowledge is to collect and process information from other people through communication with them. However, if the individuals are more than approximately five meters away from each other, the total message transmitted and sensed with our senses will rapidly decrease (Branzell 1995). This is an important reason why it is advantageous to have the project team members at least in sub-projects localized together in localities without separating walls or other obstacles.

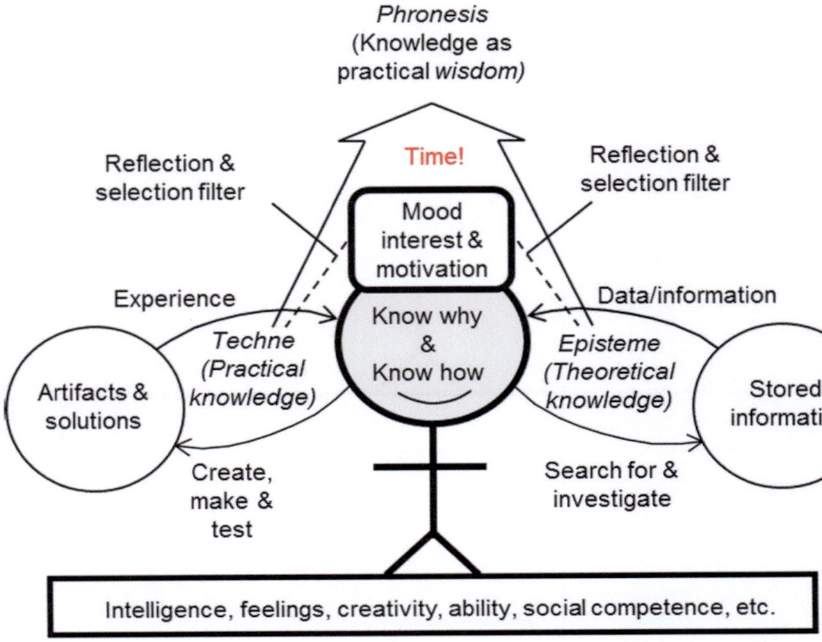

Fig. 5.7 With time, a person grows her/his Phronesis if she/he continuously expands her/his techne and episteme (Ottosson 2006)

5.6 Collective Knowledge

It has been shown that the mutual knowledge between two or more people that know each other well (e.g., in a family) is greater than the sum of the knowledge of each of the individuals (Gladwell 2000). In a similar way, when two individuals interact (communicate) with each other to solve a problem and/or to develop a new solution, they will make use of their collective knowledge. Thus, the collective—or distributed—knowledge is in good cases more than the sum of the knowledge of each team member in a project, resulting in a holistic situation. In addition, distributed knowledge reflects the fact that "no one of us is as smart as all of us".

One reason why the collective knowledge will be bigger when people with different characteristics are brought together is that our "sleeping" or unconscious knowledge from the depths of our memories will be pulled forward with the help of association tracks emerging in the communication with other people. This is why, for example, the creativity method "Brain storming" can help to find new creative solutions. Another reason is that partly overlapping knowledge between individuals will form chains of knowledge to bridge knowledge gaps between the individuals.

However, the performance of the collective activities would be hindered if the actors behave in an opportunistic way and have potentially conflicting interests

(Döner 2016). In such cases, the actors need some coordination mechanisms for synchronizing the activities and need to be able to solve conflicts for the continuation of positive interactions.

5.7 Transfer of Knowledge

The transfer of information and knowledge from one person to another can be made in different ways. One-way communication—which is used when a teacher or a speaker tells and shows something to a class or an audience—generally means that the understanding will be low by the receiving individuals. Two-ways communication—dialogues—will increase the understanding. If the teacher also checks up on the students' understanding, the situation is improved. If the person practices the new knowledge, understanding will increase even more at the same time as her/his skills improve. An old Chinese saying illustrates this:

- Teach me—and I will forget
- Show me—and I will remember
- Involve me—and I will understand
- Take one step back—and I will act

However, the last step works well only if the student is interested, motivated and in a good mood (see Fig. 5.5).

In industry, it has been popular for some time to build up computer data base "expert systems" in which experts describe their best practices. Unfortunately, the development of such systems has shown to be less valuable than the intentions had anticipated. One reason is that expert systems are seen as giving old information (Aurisicchio et al. 2006). Also, the way they were created is often not the way someone else will try to get information. Therefore, especially younger and less skilled product developers often want to consult experienced people to get up-to-date information. In addition, such dialogues mean that they will get answers and input on questions they have not foreseen when they realized they needed knowledge input. Based on this knowledge, some companies have, instead of building expert systems, built databases with knowable people (experts to maestros) in different areas that people can contact from inside or outside the organization.

Another way to increase knowledge and insights is to take part in arenas for communicative cooperation (Farner 2008). Such arenas (e.g., workshops, seminars, and conferences) facilitate possible constructive interaction between the individuals taking part in them, which can help provide input for solving a problem. In addition, taking part in arenas for communicative cooperation means automatic possibilities to meet with people otherwise difficult to reach. Thus, for example, a Ph.D. student can, at a scientific conference, interact with top scientists otherwise more difficult to get in contact with.

If a maestro walks around in a team having dialogues with the team members, this has been shown to boost knowledge and understanding. This is one reason why "comets" (maestros) are used in the Planetary Organization. In general, a powerful tool for the comets and mentors is to use so called *storytelling* to enhance understanding. Stories about failures and stories about why something did not or did work well are especially good vessels to transfer knowledge and to start dialogues and discussions.

5.8 Wisdom Streams

For classical PD models, knowledge is normally documented in drawings, in PDM (Product Data Management) files, or expert systems. In general, such data and knowledge files only tell what has been found that is working well. Eventually, some notice is given when a design change was done and who did it, providing the possibility to contact that person for further information.

However, we need a more complete way of storing knowledge that is accessible by other people in the company than that which PDM files and expert systems can offer. In particular, we also need documentation of the experience of that which did not work well, as well as ideas and solutions that have not been developed to a final product. Thus, we need to document practical knowledge (techne), as well as theoretical knowledge (episteme). When a clear picture has been created of the findings, practical wisdom (phronesis) should also be documented.

Toyota uses documentation on European A3 sized paper (also called K-Briefs, short for knowledge briefs) to document their findings from product development of new car models. The documentation on the A3 s must be condensed and easy to read and understand. Therefore, they contain curves—so called trade off curves—and sketches/drawings. Such material is an efficient way to inform about what has been found in the development (e.g., Ward 2007).

For innovation management, not only technical know-how and know-why are important to document. It is also important to evaluate and store extended information of findings in ways that many can access. We call this "corporate wisdom" and the way this wisdom is created from different innovation projects in an organization is shown in Fig. 5.8.

If experience and knowledge gained in one project is transferred to another project in an efficient way, that team can start at a higher knowledge level than if that situation does not exist. Classical PD models do not discuss the transfer of knowledge and experience between projects although this is an important issue.

Often the documentation of gained knowledge and experience is done some time after they have been made. If a long period of time elapses between when they have been gained and when the documentation is done, it is easy to forget why certain solutions were tested and why they did not work properly. Therefore, it can be advantageous to let someone else, other than those who have done the tests do the documentation.

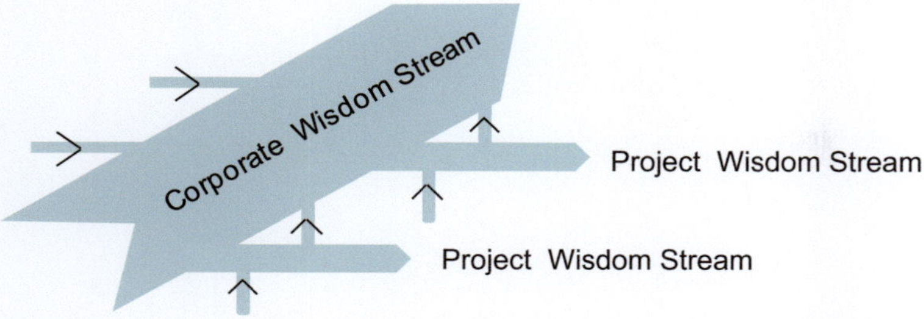

Fig. 5.8 It is important is to find ways for wisdom gains in the different projects and add them to the corporate wisdom

To be able to get extended information afterwards, it is important to make notes when the database was updated and by whom. It is also important to decide who has the authority to carry out the updating.

Chapter 6
Different New Product Development Models

6.1 Introduction

Independent of whether the development of a new product and/or service is to be done in the private sector, the public sector or in the idealistic sector, the provider faces increasing regulatory pressure and competition. Therefore, developing new products has become even more critical than in the past. For this reason, streamlining product development (PD) processes is becoming a major priority for providers looking to improve their market share, and to build responsible and sustainable activities.

Many New Product Development (NPD) models exist for *need-based* PD. *Want-* and *wish-based* development is strongly connected with innovation, but so far, these are only treated in the DPD™ method.

In this chapter some major PD methods are presented briefly.

6.2 The Classic View Versus the Dynamic View

In larger western corporations and in PD literature, there is a view that without an existing customer/product or a market need, no PD project will be successful. In general, the "need" is seen as an existing need or problem that can clearly be stated. Historically, the engineering design pioneers Fredy Olsson from Sweden and Vladimir Hubka from Hungary also included a wish, a dream or a market want under the term "need". However, when a term covers everything, it does not support the development of theory and methods.

As was briefly pointed out earlier, we have found it meaningful to distinguish between a present need, a want (i.e., a near future need) and a wish (i.e., a distant future need). Incremental innovations are often based on satisfying a want, while radical innovations often are based on satisfying a wish. The relevance of using the

S. Ottosson, *Developing and Managing Innovation in a Fast Changing and Complex World*, https://doi.org/10.1007/978-3-319-94045-8_6

three terms—and not only a need for everything—is that the conditions for want- and wish-based NPD differ a great deal from need-based NPD.

From a traditional point of view, the start of an NPD process starts with finding out "customer needs"—or simply a "need". The next step, when a need is to be satisfied, is to define the important demands for the development. In the end, hopefully, the result will be a new product that differs both in features and style from existing products. As discussed, need-based development builds on the principle that the most important demands are known or can be found before the development starts.

Need-based development is planned to be done step-by-step so that the whole development process can be broken down in chunks, often called stages or phases, with controlling "gates" separating them. Within each stage, development iterations can or are done, but not between stages. Therefore, in principle, when a gate is passed it is not possible to go back to a previous stage. In such cases, the project is normally closed and a new one opened to allow a new start that builds on the previous experiences. However, this is an expensive and inefficient way to work, for *want-* and *wish-based* development iterations often have to be done, sometimes all the way back to the start of the project. Figure 6.1 shows this basic principle with the stage-gate system principle and the dynamic (iterative) development principle needed to satisfy a want or a wish. Gates are for want- and wish-driven development rather meaningless as "milestones" reached can be lost immediately after a gate. Note that need-based development can be done without the stage-gate-principle so that the dynamic way of working also works well for need-based PD.

Sometimes the serial development model in Fig. 6.1 is described as traveling on a road with traffic lights (see Fig. 6.2), while the dynamic model is more like traveling on a road with traffic islands instead of traffic lights. The traffic islands have the effect that the medium speed will be higher. The big advantage, however, is that it is easy to drive back and forth when needed, which is a flexible and more

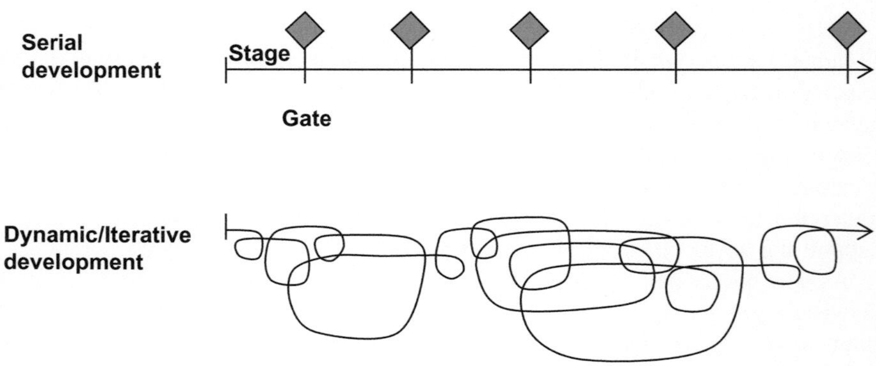

Fig. 6.1 Serial development used when satisfying a need, compared with the iterative development needed when satisfying a want or a wish

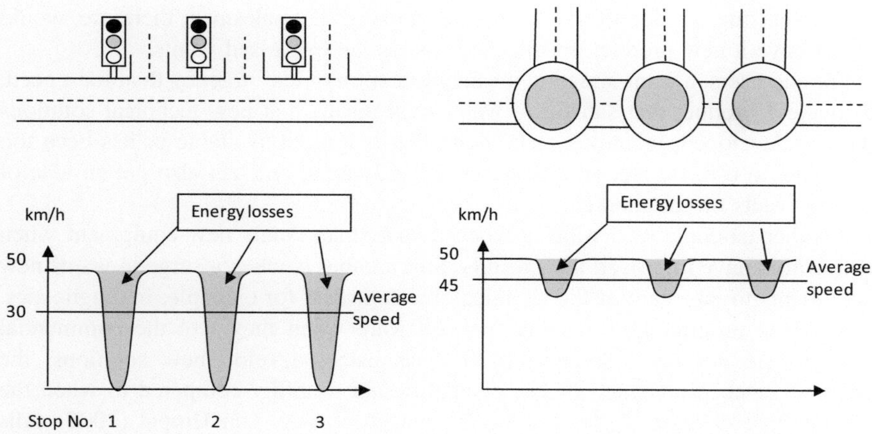

Fig. 6.2 Gates in the serial development models are like a traffic light, while in the dynamic model they are exchanged with traffic islands, meaning a safer and more efficient system

forgiving alternative than the traffic light alternative. The energy consumption is also lower, as a complete stop for red light means that "invested" energy gets lost and that more energy must be invested to start and accelerate after the stop than is the case at traffic islands. Complete stops are rare in low traffic intensity situations and acceleration will be more moderate in and after the roundabouts. The traffic island system is an example of a complex adaptive system because the drivers will adapt to the actual traffic situation by making mutual decisions.

The reason for the two completely different ways of working shown in Figs. 6.1 and 6.2 is that need-based development is based on stable conditions, while wish- and want-based development is based on unstable conditions. While the target is well defined and stable for need-based development, the target for want-based development is moving, together with the fact that many unknown variables exist in that case. For wish-based development, the target is a vision, which means that it is rather impossible and meaningless to plan carefully for the long term. Instead, the development team has to create solutions, make models and prototypes and test them to learn how to make better solutions. The more tests per time unit, the faster and safer the development will be.

For want-based development, the ability to adapt to the changing situations is important. Therefore, in short, the big difference between need-based classical development and wish-based dynamic development is that the customer will get a solution that is right when she/he gets it, while she/he gets a product that is in accordance with what was decided when the development started some months and even years ago when a classical view is used. The want-situation is more like the wish-situation than the need-situation.

A want-situation for an organization is, for example, when a salesman experiences that some users would benefit from having a product that has other features than the existing products on offer by the company. The salesman, therefore, would like to have a new product to sell that satisfies the expressed wants.

For end users, a want-situation means a situation of less urgency than for a need. To make a product that satisfies a want often means that new technical solutions must be developed, including technology that is not yet available or has been too expensive to use. By studying trends and trendsetters, one can also get an idea of coming wants on the market.

Another example of a want is when a sportsman wants new equipment when competitors have improved their results. Still another is when a surgeon wants new instruments to take care of the technical development, for example, in diagnostics. A fourth is programmers making new software when they feel the commercial products are not good enough. If they, as users, develop new solutions, the solutions reached are often of higher quality and usability compared to when the solutions are developed in a traditional, industrial way. von Hippel (2005) calls these user inventors the *lead users*.

Thus, when a want-situation arises, one generally has some time to develop a solution, which also is needed as the possibility to copy already existing solutions (even called *reversed engineering*) or parts of solutions is normally much lesser for want-situations than for need-situations. Therefore, a want-situation requires a higher content of product design/innovative development than re-engineering does. As a consequence, for want-driven activities, it does not make sense to make long-term detailed plans to reach the target because many unknown problems have to be solved along the way and these can destroy thoroughly developed plans. However, short term plans and rough long term plans are important to do, not least to learn from reflecting on the deviations from the plans.

A wish-situation is at hand when someone expresses a wish for a visionary product or solution. By asking people what they would like to have if there were no technical or economical limitations, a company can get some ideas on which to base further work. Another input for a company to get a feeling for what can be a new market possibility is to study trend setters and trends. Investigating new scientific breakthroughs in ground research (e.g., physics, chemistry and mathematics) and/or more applied research fields, such as medicine, bioscience and ICT (Information and Communication Technology), which can also bring ideas for new products. Investigating patents is still another way of finding ideas.

In general, when a wish-situation is at hand, one has some time to develop product solutions. However, it is often quite impossible to predict when a solution will be ready as there are few existing solutions to copy or use. Thus, an open end of the project is at hand, which is outside the classical definition of a project, which says that a project should have a clear limit of Cost, Time and Performance. The open end situation for wish-based development is sometimes disturbed, for example, by politicians and top management of larger companies who decide when new solutions must be ready.

Before going into details of PD models, a discussion of research on PD models is done, as the reliability of the findings from the research is based on how the data was collected and who was interpreting the information.

6.3 Research Methods on Innovation and PD

Individual and collective knowledge increases when new knowledge is added to existing knowledge, or when we adopt a different perspective on knowledge. As was discussed in Chap. 5, Aristotle divided knowledge into theoretical-scientific knowledge (episteme), practical/productive knowledge (techne), and knowledge of practical wisdom (phronesis) (Gustavsson 2000). Techne and Phronesis are present and interlinked in a complex pattern in product development.

Consequently, research on development processes is done to extend knowledge and to give input for new theories, and/or to find out how well theories function in real processes. Such research can be pursued in the form of prospective studies in real time, or as retrospective studies in past time, as two opposing possibilities (see Fig. 6.3).

For studies of ongoing processes, such as product development and innovation processes, it is recommended that the lion's share of the research would be prospective studies (Ottosson 2003). Further, a Phronesis research approach should focus on practice because human action cannot only be understood or judged through generalizations, static concepts or universal theories (Björk 2003). Instead, it is found in practical knowledge that builds on human experiences (Gillberg 1999). Therefore, we must study practice as the concrete particulars of reality (the priority of the particular), which is complex and constantly changing.

In general, own use and own participation by a researcher means that the reliability gets higher than if the researcher chooses other methods, as is shown in the two left hand boxes on the middle row of Fig. 6.3. This is because the information gained in the right hand box is first-hand information, without interpretation errors. In principle, reliability is dependent on the method used, as shown in Fig. 6.4. A qualitative treatment is used in the left of the Figure, while quantitative

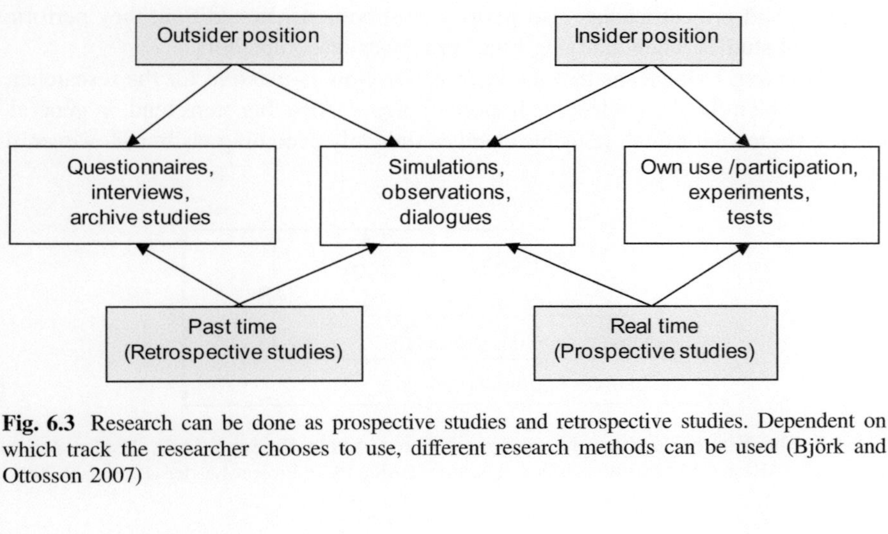

Fig. 6.3 Research can be done as prospective studies and retrospective studies. Dependent on which track the researcher chooses to use, different research methods can be used (Björk and Ottosson 2007)

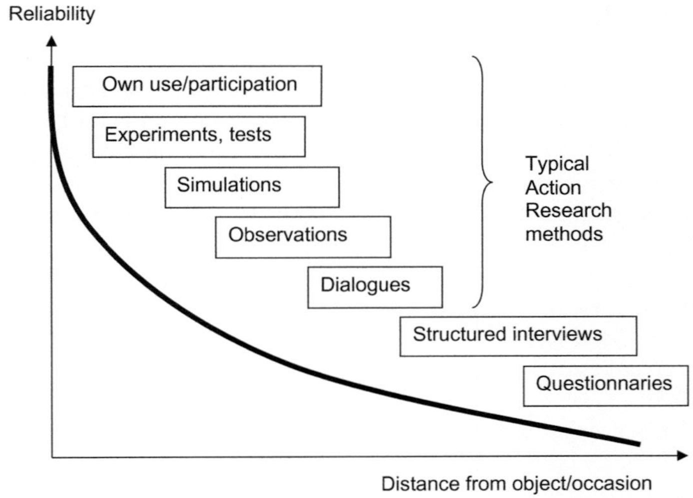

Fig. 6.4 Reliability, depending on the investigation method used

treatment is dominant in the right of the Figure. Note that the same reliability situation is at hand both for researchers investigating development processes and for product developers investigating the situation for use of a product.

Performing Action Research (AR) means to be inside the studied process, which is still not fully accepted as a "good" research approach by researchers who stick to the old view that objectivity exists, and that quantitative studies give reliable results.

AR can be performed in three ways; as an observer, a team member or as a project leader/manager (see Fig. 6.5). Anyone can, in principle, be an observer, while few researchers have the possibility to be team members in industrial processes. Still fewer can be project leaders or managers. For natural reasons, team members and project leaders also perform observations. In addition, they perform traditional studies when studying other processes or competitors.

Conducting IAR means that the information flow is massive for the researcher, and it can be difficult to select the important pieces. Also, big steps tend, in general, to have started with small invisible changes suddenly becoming visible. However, if

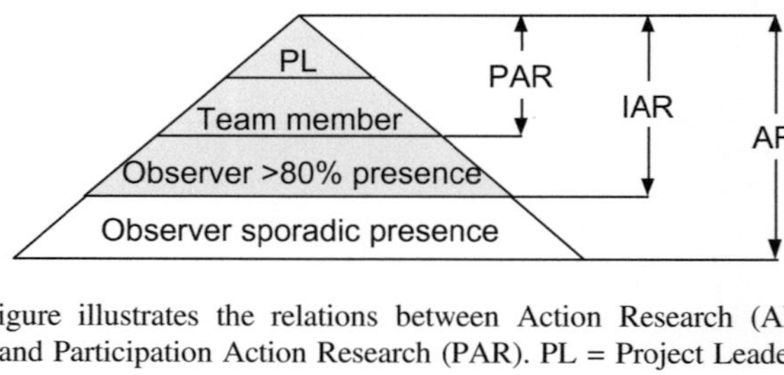

Fig. 6.5 The Figure illustrates the relations between Action Research (AR), Insider Action Research (IAR) and Participation Action Research (PAR). PL = Project Leader (Holmdahl 2007)

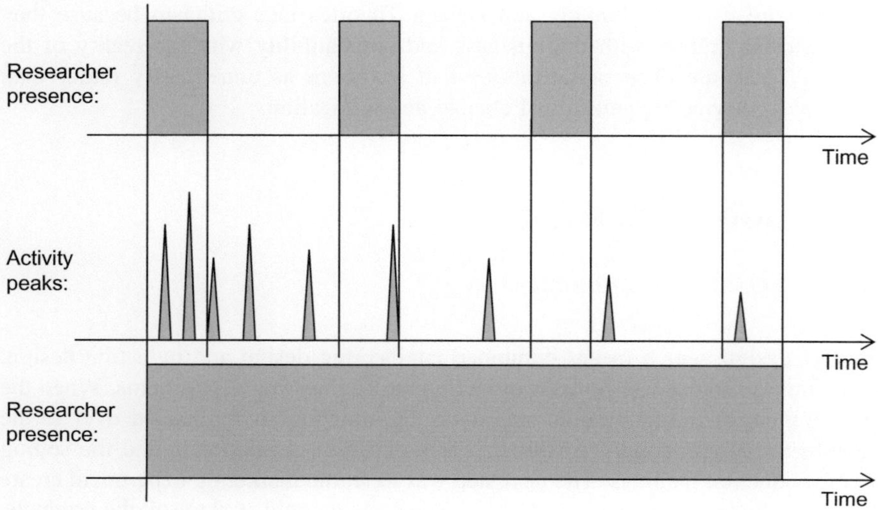

Fig. 6.6 Activity peaks of different magnitudes occur and are unevenly distributed. To catch them, it is not enough to be present now and then (Björk 2003)

much of the communication is done over the Internet, later reflections can be done by looking at the saved files.

Being present most of the time in a development process means also that the small changes, which appear unevenly distributed in time, can be grasped (see Fig. 6.6). Therefore, Insider Action Research (IAR) is recommended to give a solid understanding of such a complex process as PD and innovation development.

In principle, an observer does not need to be accepted by the people in the studied process, although practitioners can be frustrated by having someone present who does not take part in the everyday work. Thus, the duality situation should not be a big problem for observers being present most of the time in a development project—while being part of a team or being project leader of a team, and simultaneously being a researcher of the project, can be problematic for time and engagement reasons. The project leader, as well as the team members, will have a political role during the work, while an observer will not have a political role when she/he makes observations. However, observers can play an important political role when the report has been made public.

A conclusion is that research based on PAR gives the best situation for understanding the complexity of PD and innovation management, which, in turn, can be used for giving useful recommendations to practitioners.

In our studies of different innovation and PD theories and methods, we have found that the theories concerning the meaning of systems of assumptions, principles, and relationships often are weakly grounded. The use of empirical methods (e.g., experimentation, observation and experience) to obtain empirical evidence that justifies the truth or the plausibility of a theory or that falsity it are not frequently used.

Therefore, Design Thinking and Design Theories face criticism because they lack both integration with the business and compatibility with the reality of the market. There are even organizations that see them as unnecessary rather than essential to driving organizational change and innovation.

6.4 Classical NPD Models

6.4.1 *Serial Development*

Product design, which means combined engineering design and industrial design, was, until the end of last century, often done in engineering departments. When the development of a new product was done, the drawings were handed over to the manufacturing/production department for production development and the setting up of production facilities. The next step was to let the marketing department create marketing material, after which the sales department could start to sell the products, ending up in new innovations.

This serial/sequential model is often described as an "over-the-wall" model, as there are notional high walls between each department, over which to "throw" the work to the next department when finished (see Fig. 6.7). It is also called the "relay race method" because the "baton" is first carried by one department, then by another (Rama and Herbig 1996).

The work in each stage in serial product development is governed by rules, detailed planning, long detailed specifications, and reorganizations. The walls between departments have a striking resemblance to the gates in the Stage-Gate® method described below.

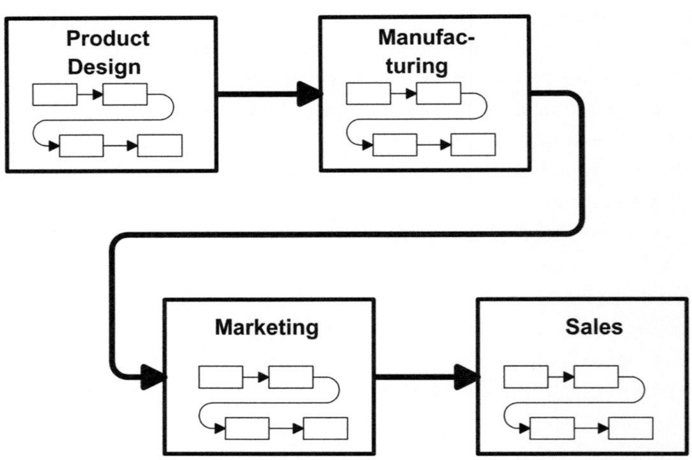

Fig. 6.7 The serial development model for hardware products is still used in some companies

Unfortunately, the serial way of working often leads to products that do not fulfill the customers' needs and demands. It is also inefficient, costly, and sometimes results in poor quality products (Ullman 2003). Although these and a number of other negative aspects are known, many companies are still using serial product development.

One reason that serial development is used is that management does not want to let go of detailed plans and micromanagement (even from afar), as they give a feeling—or an illusion—of control. Another important reason is that the buildings, through their physical layout, often support or enforce serial development if they are not re-built.

6.4.2 *Stage-Gate® (SG)*

The Stage-Gate® (SG) model for product development has been popularized by Dr. Robert G. Cooper (originally at McMaster University in Canada). Cooper's model is a formalization of general task management principles that were created by NASA decades ago for managing massive aerospace projects.

Thus, in 1986, Dr. Cooper presented a product development management model which, from 1988, he called the Stage-Gate® model (see Fig. 6.8). In the third-generation, 2001 edition of Dr. Cooper's book, "Winning at New Products", a "discovery stage" comes before gate 1, and a "post launch review" comes after stage 5 (Cooper 2001). Now "Fuzzy gates" are also allowed, permitting some overlap between two stages, as in the Waterfall model.

SG builds on information from a large number of companies that have told the researchers about the best ways of working, which has then been used to formulate best practices. It has a stronger focus on the decision points between each main activity than the other classical methods so far discussed. The decision points are called "Gates" and the actions between the gates are called "Stages". SG does not prescribe what to do in the stages between gates.

For Stage-Gate®—as well as for the other classical methods—the ambition is to filter out different ideas and solutions as early as possible, and to concentrate all resources on developing one solution. This solution should specifically meet set

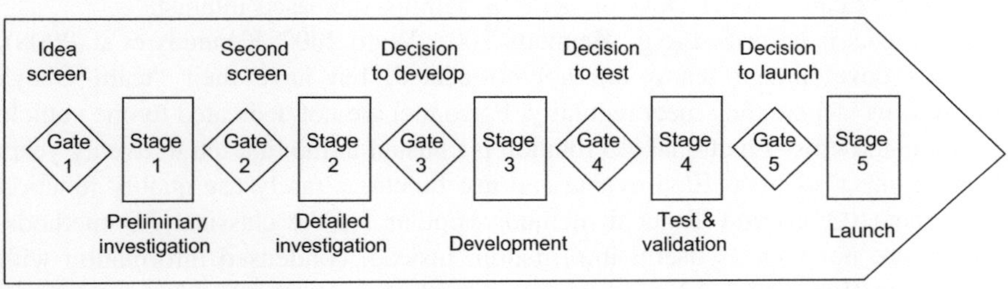

Fig. 6.8 The Stage-Gate® model (Cooper 1994)

demands on function, development cost and development time. At the gates, decisions are taken either to "Kill" the project or to give it a "Go" to the next stage. As the gates are critical, the presentations by the project leaders must be well prepared. Therefore, in larger enterprises and for their larger projects there is a saying that "all development work stops three weeks prior to a gate review", as everyone focuses on preparing for the meeting.

Stage-gate models demand strongly linear proceedings. That is, officially the next project stage can only be initiated when all tasks of the preceding stage are completed and positively evaluated (Lindeman and Lorenz 2008, p. 175). Hence, radical innovation projects as well as hardly predictable ventures, where neither a competitive environment nor a solid customer base is known at the beginning, will find little support in stage-gate models (Lindeman and Lorenz 2008, p. 175).

SG is supposed to facilitate the low-risk undertaking of product development, especially new product development, which is classically where risk is highest. Unfortunately, SG methods lead to bureaucracy, a focus on early "killing" of product ideas that in changed circumstances could be winners, and delays when time-to-market should be paramount for need-based PD.

6.5 Dynamic NPD Models

6.5.1 Lean Product Development (LPD)

The car maker, Toyota, is commonly regarded as superior to other larger car makers on the basis of different, measurable criteria. The origins of "lean"—which basically means avoiding all types of waste (e.g., Johnstad et al. 2012; Womack et al. 1990)—are found in the Toyota Production System (TPS). Toyota would say: keep it simple, trust your employees, make sure everything is visible to them, and enable them to act on what they see. Toyota is proof that high quality results can be consistently delivered on time, with minimal process overhead. Therefore, the Toyota philosophy in the form of "lean" principles has spread, for example, to the health care systems, independent of whether they are in the private or public sector.

Also, Toyota's product development process for new car models is known to be excellent, as it provides short time-to-market compared to other car makers. Therefore, "Lean Product Development" is gaining increased interest.

From what is described (e.g., Kennedy 2003; Ward 2007; Kennedy et al. 2008), Toyota's development teams are not co-located, but have their "team Obeya (war) rooms" where they meet regularly. Personnel are not dedicated to one vehicle program, and cross-functional job rotation is unusual in the first ten to twenty years of an engineer's career. Engineering and test functions rarely use quality function deployment (QFD) and Taguchi methods popular for the classical PD methods. Reports are not seen as useful information. Instead, condensed information with sketches on (European) A3-sized papers is used.

At Toyota, the view is that "an engineer should never be more than a stone's throw from the physical product". This principle is referred to as "Gentchi Genbutsu". It is practiced in many ways. Examples include spending a significant amount of pre-program time at manufacturing plants and dealerships, working on competitor tear-downs, or personally fitting parts on prototypes.

Entrepreneurial leadership is a must in LPD, meaning that the Chief Engineers—or Entrepreneurial System Designers—use mentoring leadership and actively move around: "Do not rely on reports but go see it yourself!" Mentoring leadership means that the managers do not lead by giving orders, but by asking questions. The Chief Engineers are very experienced engineering managers who have been trained to be customer experts. The customers of the product developers' work are defined as end users, manufacturing, sales, or agencies.

Instead of turning the focus to one solution as early as possible (called Point-Based or Single-Event Concurrent Engineering), many possibilities are worked on in parallel in order to bring forward many workable solutions (called Set-Based Concurrent Engineering-SBCE). Thus, SBCE is the shift from developing and testing the design for a particular project to testing, learning, and considering a larger set of possible designs. In short, Toyota organizes and develops product capabilities from the bottom up.

As LPD is need-based and the PD process starts with what is known from past projects and the identification of what one needs to learn first in the new project. It then proceeds to learn what is needed to make decisions. Specifications ('specs') emerge along with the design, based on what has been learned about the customers, the products and the technologies the products are based upon. Instead of listing specs at the beginning of projects, targets are set that express sets or ranges of customer interests that one wants the product to end up with, though one may not yet be sure which of these are possible, or compatible with each other. "Expert Engineering Workforces" are set up for the development of new cars. They have their own responsibility for the planning and control, as no planning departments exist at Toyota (Ward 2007). The internal planning and control in each group are visualized in the Obeya rooms, for example, by writing on the walls.

Decisions are taken as late as possible, so as not to limit flexibility. This includes avoiding early concentration on one best solution to develop to a finished product. All developed solutions are saved for later use. Eventually, parts from the different solutions are combined to build the first final solution. So called "trade-off curves" are used to find the best individual solutions, which are comparisons of a number of features of the various solutions.

Product Development is seen as a learning process and knowledge gaps must be removed before the design takes place. 'LAMDA'—Look, Ask, Model, Discuss, and Act—is important behavior to adopt during the learning process (Ward 2007). What has been learned is continuously documented on A3 s in an informal and easily understandable style. "If the knowledge is not visual and easy to understand, it's virtually useless." A3 s are a living part of the knowledge flow, not a historical reference. If computer documentation is used, it needs to maintain the simplicity of a single page A3 and provide the ability for the reader to follow the LAMDA storyline.

It is not revealed in the literature whether these A3 s are scanned and stored on Product Data Management (PDM) or Engineering Design Management (EDM) files, or how, in such cases, they would be made searchable.

The A3s successively build up knowledge standards called "Knowledge Briefs", "K-Briefs" or "Knowledge check sheets". These K-Briefs are used in other product development projects to prevent knowledge waste, and to prevent the repetition of mistakes. Knowledge gates are not used as task gates, because they are in the stage-gate methods. Knowledge databases are questioned. This is because the most important learning often comes from what did not work, which is not contained in traditional databases. Thus, no stage-gate systems are used that expel control in the gates.

Critics of LPD say that the messengers of the method do not really know what Toyota does. "After thirty years, we can now be reasonably certain that whatever Toyota have got, it isn't a trivial task to bottle it and sell it on." (New 2007). Further, the A3 documentation principle seems problematic in the modern Internet society. Finally, the philosophies behind the model are not anchored in science. The often used two stream principle for LPD shown in Fig. 6.9 is an example that "Feedback" is seen as a pump, pumping back knowledge from one "creak" to the main "river" from which the original knowledge came.

Note that the term "lean" is not protected, which is why there are as many interpretations of what lean ways of working are as there are writers, and everybody has her/his own view.

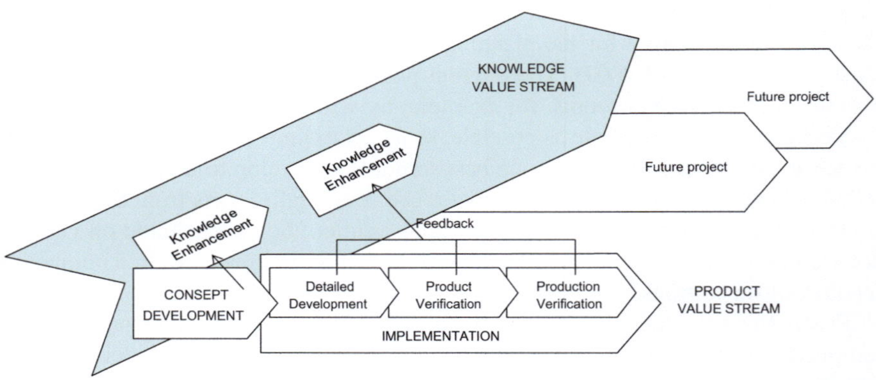

Fig. 6.9 The lean product development process as it often is described (Swan and Furuhjelm 2010)

6.5.2 Agile Software Development—ASD

Agile Software Development—ASD—refers to a group of software development methodologies that are based on similar principles, and which have been developed since the 1990s. Agile methodologies generally promote a project management process that encourages (e.g., Highsmith 2004):

- frequent inspection and adaptation; a leadership philosophy that encourages team work, self-organization and accountability
- a set of engineering best practices that allow for rapid delivery of high-quality software
- a business approach that aligns development with customer needs and company goals

Agile methodologies, such as Adaptive Software Development, SCRUM or Scrum, eXtreme Programming (XP), Feature-Driven Development (FDD), and Crystal, all strive to reduce the cost of change throughout the software development process. For example, XP uses rapid iterative planning and development cycles in order to force trade-offs and deliver the highest value features as early as possible. In addition, the constant, systemic testing that is part of XP ensures high quality through early defect detection and resolution.

Behind the development of the ASD concept was the insight that often, when a solution has been developed according to classical ways of working and is implemented, the environment and requirements have already changed, calling for expensive changes. ASD emerged as an alternative to document-driven, rigorous software development processes (Highsmith 2002). Software developers realized that development processes that also require many documents, artifacts and procedures are too slow to fulfill customer needs. Moreover, nowadays, business needs change faster than software projects following old methods are able to keep up with. Therefore, the focus had to switch from fulfilling predefined project requirements to delivering up-to-date value to the customer.

Agile software development also aims to minimize failure risks by developing software in short periods of time. Software developed during one unit of time is referred to as an iteration, which may last from one to four weeks. Each iteration is an entire software project, including planning, requirements analysis, design, coding, testing, and documentation. An iteration may not add enough functionality to warrant releasing the product to the market, but the goal is to have an available release (without bugs) at the end of each iteration. At that time, the team re-evaluates the project's priorities.

Agile methods emphasize face-to-face (F2F) communication over written documents. Most agile teams are located in a single open office, sometimes referred to as a "bullpen". At a minimum, this includes programmers and their "customers" (these customers define the product; they may be product managers, business analysts, or the clients themselves). The office may include testers, interaction designers, technical writers, and managers.

Agile methods emphasize working software as the primary measure of progress. Because of the preference for F2F communication, agile methods produce very little written documentation relative to other methods. This has resulted in criticism of agile methods as being undisciplined and, in fact, the approach is not grounded in science.

6.5.3 Agile Hardware Development

In 2016, scientists in Germany (Schmidt and Paetzold 2016) concluded that agile methods are serious alternatives to traditional approaches for physical product development as "Product development becomes increasingly uncertain and dynamic. Under these conditions, traditional approaches, such as VDI 2221 and VDI 2206 reach their limits as they are too stiff to handle a large number of changes efficiently. Agile methods, however, are flexible and embrace changes."

In a follow up paper, the importance of more reliable customer feed-back to the product development teams is discussed and the term 'user experience (UX)' is introduced (Schmidt et al. 2017)

6.5.4 Dynamic Product Development—DPD™

To cope with both increasing complexity and unstable conditions—which the classical models were not developed to do, the Dynamic Product Development (DPD™) model has been under development since 1995, mainly in Sweden. DPD™ is a holistic philosophy simultaneously satisfying relevant demands on user/use, society, and business (USB). This is independent of whether the aim of the organization is a better world, a good service or a sustainable profit.

To ensure clear understanding of the DPD™ model, Ottosson and Partners (www.ottosson.biz) was established in 2008 to avoid facing the same problems with different meanings of the term as has been experienced with IPD and Lean Product Development.

The next chapter describes DPD™ in more detail. However, one very special and important rule is not to list many demands and then to solve them as for the classic start (see the upper part of Fig. 6.10). Instead, there should be only a few demands to solve when the appointment of the project team leader takes place. It is also important that the project leader has the responsibility for appointing her/his team and not the reverse, which was the classic way of working.

Working in the dynamic way reduces the time to a ready product considerably, as is shown in Fig. 6.11 shown as T/N. It is not uncommon that the number of demands to solve in the traditional way of working can be 100. The number of remaining demands to solve for the dynamic way of working should not be more than four at any time in the development process.

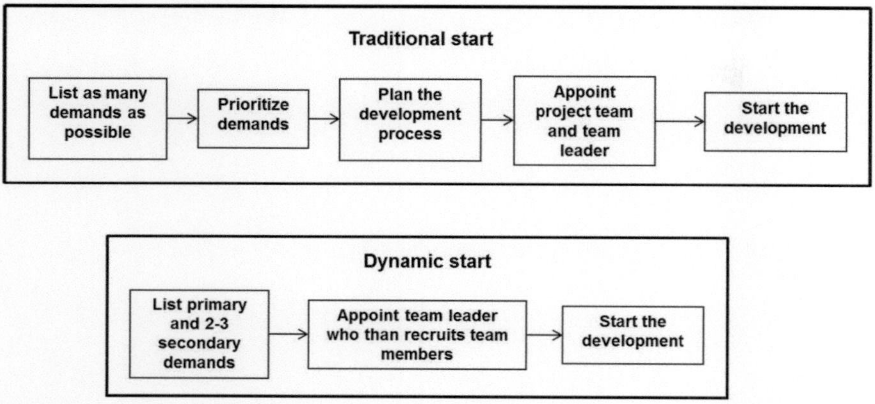

Fig. 6.10 What precedes the start of a traditional product development project differs much between the traditional and dynamic views

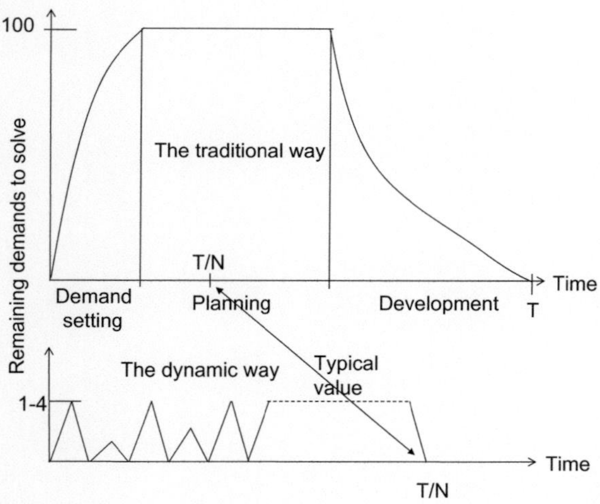

Fig. 6.11 An important difference between the traditional way and the dynamic way of working is the number of demands to solve at each moment of time

Some reasons why this way of working results in a shorter development time are:

- Human beings can handle 3 ± 2 items simultaneously.
- It is difficult to plan many items.
- It is easier to focus all energy on few items than on many.
- It is easier to motivate people when they have small steps to overcome.
- Fewer demands to solve means less complexity.
- The situation is less stressful.

Chapter 7
Dynamic Product Development (DPD™)

7.1 Introduction

This chapter is about how to develop new products. However, the principles can also, to a great extent, be used for the development of marketing material, organizational development, and personal development.

For an innovation project to be successful, it needs as fast as possible to work concurrently at product and market development. In this process, it is beneficial to sell the new product before it is completely developed, as bringing in a demanding customer and/or user means making the project real, something that can never be accomplished in laboratories. It also constitutes a positive psychological attitude for the development team, the financers, and the shareholders as they get confirmation that there is a real market for the product even if, initially, it is to be sold at a low price. However, it is difficult to sell something that cannot be shown, which is why it is beneficial to have a functional prototype to show.

Generally seen, Product Development (PD) is a learning process driven by the need to take care of new opportunities or an existing or constructed need, want, or wish. Table 3.1 shows the main differences between these three sets of starting conditions. As described, need-based PD projects have stable conditions to work with, while want/wish-based PD projects experience unstable conditions. As discussed, two philosophically different views exist on how to best perform need-based PD development, leading to a categorization of PD methods as classic and dynamic, depending on their ability to handle stable/unstable conditions.

As was discussed in Chap. 6, the classic PD models all presuppose an ordered world for which a good result can be achieved through detailed planning. The dynamic models, such as Dynamic Product Development (DPD™), Lean Product Development (LPD) and Agile Product Development (APD), are designed to handle unstable conditions and the increasingly complex developments typical of New Product Development (NPD) projects, where meaning and patterns are often only discernable in hindsight.

S. Ottosson, *Developing and Managing Innovation in a Fast Changing and Complex World*, https://doi.org/10.1007/978-3-319-94045-8_7

The three dynamic models have much in common although DPD™ has a stronger, more evolved and effective control structure than other methods (Holmdahl 2007). LPD is designed for need-based hardware, PD, and APD was initially designed for software development. DPD™ was originally designed for want- and wish-based PD for all types of products. However, most traditional models are now incorporating dynamic/agile parts in their principles (Ottosson and Kolla 2017).

It is difficult to find out which PD method is the "best" from comparing studies as there are so many things that will influence the result and influence the functionality, the time it takes to develop the products, and the cost of the development. However, in a study comparing the development of two similar new stairway wheelchair lifts for the same use, a time-to-market difference of about ten times showed up between a commonly used PD model (Concurrent Engineering) and DPD™, to the advantage of DPD™ (Ottosson 1996). For the software development of similar products, when a classical method (RUP®) and DPD™ were used by two similar teams, the study showed that time-to-sales was ten times shorter for DPD™, and three times shorter for the fully developed DPD™ product (Ottosson 2004).

7.2 Organization

When a company or an individual has decided to develop a new product as part of a new innovation, the ideal situation is to first engage an entrepreneur or an intrapreneur who, in turn, can appoint a PD project leader if she/he does not take on that duty. The project leader will have as a mission not only to ensure that the new product is developed, but also that it can be transformed into a successful innovation. Therefore, it is essential to have an entrepreneurial project leader who, if possible, also has some market experience and knowledge. This is independent of whether the development is to be in the private, the public or the idealistic sector.

The startup for the PD project leader is to find, with the help of inventive people, a number of concepts that will solve the need, want or wish that the project has to satisfy. If, for example, after a first brainstorming session the entrepreneurial project leader gets three interesting concept suggestions and two more after the second brainstorming (see the upper part of Fig. 7.1), she/he can, if there is money for that, ask five (experienced) developers to materialize each of the five concept suggestions, one per person (see lower part of Fig. 7.1 where E stands for entrepreneurial project leader and the numbers refer to each developer). The developers have to report their findings to the entrepreneur or intrapreneur, who should also take an active part in the progress. This is accomplished by using Management by Walking Around, as discussed in Chap. 3.

After some time, the PD project leader and entrepreneur/intrapreneur in the example will find that concept 1 seems not to be worth going further with. With new knowledge, however, this concept can be opened again. After some time, the intrapreneur finds out that the concepts 2 and 3 can be merged to be concept 6.

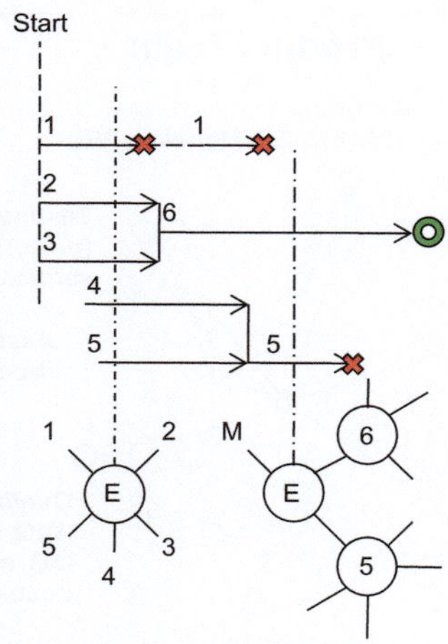

Fig. 7.1 The principle of developing different concepts into one new product

In this example, concept 4 can be added to concept 5 instead of being developed further on its own. Now it is apparent that concept 1 does not hold and only concepts 5 and 6 will remain for further development. At this time, more people are needed for these two concepts so team 6 in Fig. 7.1 will have a team leader and three team members. Team 5 will have one team leader and four team members. Gradually, however, it turns out that all concentration will be put on concept 6, which is why more resources have to be allocated to the two remaining project teams. M in the Figure stands for marketer, which means that one marketer is involved when the two promising concepts have been found.

Thus, in DPD™ the work to develop a new product starts with the creation of several possible concepts and through a process of continuous testing and evaluation arrives at the solutions and/or concepts that will be finally developed. By refining, combining and deleting them, the number of concepts will gradually decrease until only one is left.

The work on detailed solutions is generally time consuming, involving many work hours as it must be done carefully and with great responsibly. It is a known fact that even a small mistake at the detail stage can be catastrophic for the whole product. "The devil is in the detail!" is an expression that highlights this reality. When problems occur on the detail level, one may have to go back to the abstract level to find solutions from which to work further.

According to our experience, especially when developing a new product for which no or few known solutions exist or when problems occur in more traditional product development processes, the classic line organization gives a management situation that is too weak to get an acceptable result, measured in time, cost and performance of the project. In particular, commonly used line and matrix organizations have the disadvantages of reacting too slowly to changes. This means extra costs compared to when an organizational method is used that has not been built in

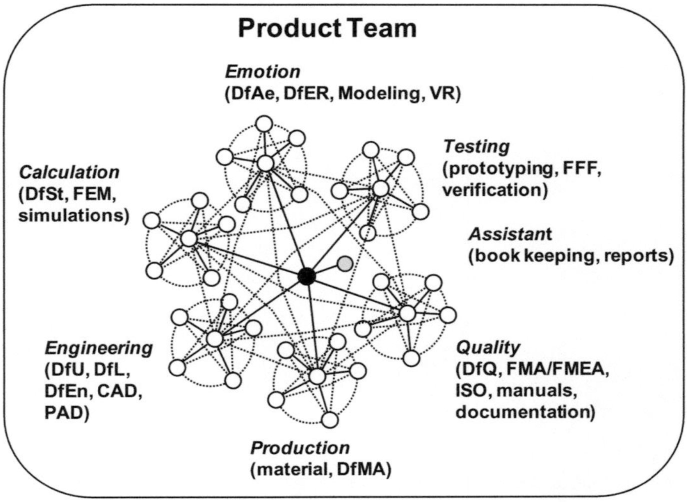

Fig. 7.2 Different topics to cover when a complicated product is developed (the terms will be discussed later) (Ottosson 2015)

such negative elements. Therefore, the planetary organization explained in Chap. 4 is used in DPD™.

Using the planetary organizational form for the development of a complicated hardware product (such as a new car), the organization can be set up as is shown in Fig. 7.2. However, one basic principle of DPD™ is to have organization that is as small as possible all the time and manning that is be flexible at each point of time. Therefore, positions are seldom occupied all the time by one person even late in the development, and thus, close to the delivery of a complete product. One person can also be responsible for different positions. In the Figure, abbreviations are used that are explained in the remaining text.

With increased numbers of people that need to be engaged in the project, when a concrete and detail level has been reached, more space will also be needed.

The longer the project runs and the more work that is done at the detail level, the more natural it will be to split up the project into sub-projects. However, without close contacts between the team members within each sub-project and between each sub-project, the work will be less efficient. Thus, if different sub-projects are placed close to each other, based on the need for mutual contacts, a more favorable situation will be reached than if sub-teams are placed in localities apart from each other.

The project and sub-project leaders must be movable within and outside their teams in order to be able to be one step ahead of the team members. Therefore, the principle of Management by Walking Around is used to make it possible for them to experience, with all their senses, the actual situation in their projects instead of relying on formal reports.

In DPD™, early preliminary and late final decisions are taken, which is contrary to the classical PD models discussed in Chap. 6. This principle means that the risk decreases that no (commercial) product will be the result after a first brainstorming session of the development at the delivery date, which is opposite to the situation when all efforts are concentrated on one early, picked-out concept.

7.3 Users/Use–Society–Business

For the PD project leader and each team member to successfully develop products and especially new products, some good advice would be to keep in mind three outer and three inner demands. These outer demands can be called: User/use, Society and Business—the USB demands (see Fig. 7.3). The Users of the products will experience the products and find out their usefulness. The product will not violate the rules and laws of Society. With Business is meant that the company—or the non-profit organization—will benefit from the work being done in some way.

As Fig. 7.3 indicates, everybody taking part in the development will be guided by her/his view of life, morale, and experiences. Independent of that, everybody in a PD project will strive to satisfy high demands, not only on making a profitable product for the company but also in making the users and society happy with the product. This philosophy is different to most other PD models, which often only have the business perspective in mind.

DPD™ is a User-centered Design (UCD) model. This means, for example, that the solutions will be functional, with good semantics to create good comfort and pleasure in the use of the products.

Product semantics is an attempt to convey what a product is or does through its form. It is based on the idea that a product's form should readily communicate the function of the product. Thus, a product with good product semantics would typically not require the user to read a manual before beginning to use it.

However, in real life, customers are often not the users or the consumers. As discussed earlier, the main difference between users and consumers is that users use

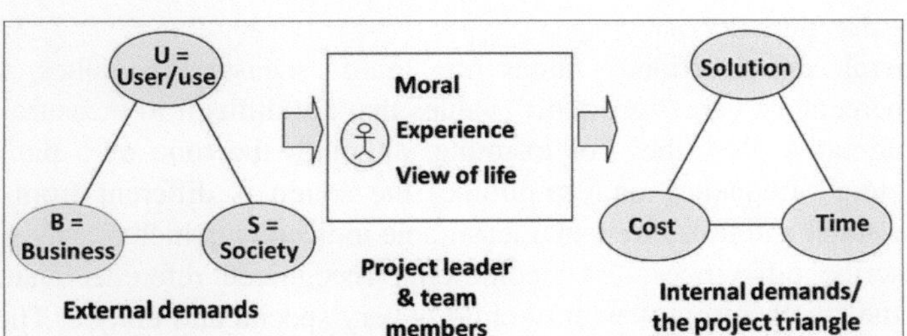

Fig. 7.3 Outer and inner demands for each team member dealing with technical questions of the product development (Ottosson 2015)

a product over and over again, while the consumers benefit from the product once or a few times. For example, consumers benefit only once from food or drinks but can benefit many times from watching a movie or listening to a piece of music. When developing a new product as part of an innovation project, the product developers have to concentrate mainly on the users, while the sales and marketing people have to focus on the customers.

However, a tight connection must be maintained between the sub-project leaders in the PD project and the sales/marketing people so that, for example, the findings and experiences from the users and the customers are transmitted easily to the other team members. The special planetary organization (see Fig. 7.2) supports such connections.

Of great importance for all involved in a product development project is to always keep in mind that it is advantageous to try to find solutions for which the users can use already fixed behaviors, not only because the users will find a greater pleasure if that is the case, but it also has a definitive positive influence on the sales and profit figures.

7.4 Product Values

As was discussed briefly in Chap. 2, from a user's point of view a product can have different product values, which can overlap each other.

For example, the functional and sensorial/perceptory values can be experienced as different "quanta", such as sound, smell, heat, vibrations, or information (see Fig. 7.4) that are emitted from the product. These quanta are received by our senses in both a passive and an active way, such as when we touch the product.

Before the functional values have been developed, it is a waste of time and money to work on sensorial/perceptory values and even more so, on the image values. For the development of the functional values, which thus are mostly hidden inside the product, engineering design is the main activity. The perceptory values are often partly taken care of by industrial designers and/or by engineering designers. The image values are mostly taken care of by the *Intellectual Property* (*IP*) and marketing experts. The development of emotional values is mainly taken care of by the marketers.

In general, the functional values are "hard" measurable values, while the sensorial/perceptory values are "soft" values that are difficult to measure and more or less difficult to describe. For example, although the tone of a piano can be measured (e.g., frequency and amplitude) the sound is different from that of a violin. For smell and taste, we lack measurable indicators, which is why we have to try to describe differences using commonly recognized reference material. For example, the smell from a new born child is very special and unique. The smell of laundry that has dried in the summer breeze is another rather special and unique smell. Generally, taste seems very difficult to describe and must be connected to many special tastes. For example, the taste of a Granny Smith apple can be

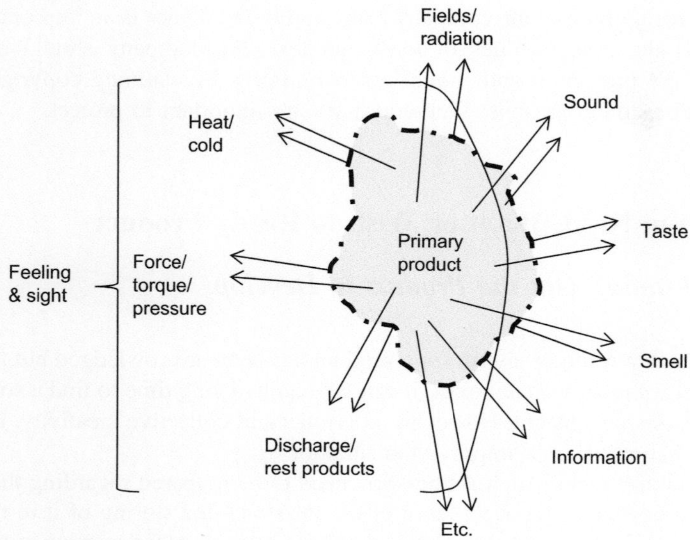

Fig. 7.4 A product emits different "manifestations" that are received by our senses

described by a trained and sensible taste tester as "acid, somewhat mealy, with a taste of light rotted tree trunk". Although temperature can be measured, the individual feeling of warmth and cold that one experiences is different and dependent on, for example, dampness and wind speed. Experienced levels of pain will differ greatly for different test persons.

A complication with soft values is also that cultural and other differences means that a product that gets high appreciation in one geographical area can get low appreciation in another area not far away. One kind of potato crisps/chips that is popular in one part of a country may not even be saleable in another part of the country. Therefore, in the global world, the parts and systems of a hardware product that make up the functional values can, in principle, be developed anywhere but to please the soft values of a local market, they should be developed locally. Thus, for product development, what is called "Glocal" (global and local) thinking becomes more and more important.

Although difficult to design and measure, soft values become increasingly important for any commercial product. For example, every premium car maker must spend a lot of effort on industrial design, sound, smell, comfort, and image values. Every fast food producer must spend a lot of effort on design, taste, smell, and image values. New occupation titles such as sound engineer, taste designer, and smell designer will therefore probably be common in product-based companies in the future.

Important image values include the brand name, recognition marks (logos) and the company name. Car manufacturers, for example, try to ensure that buyers in some way recognize new models and associate them with the brand names and traditions of the earlier generations of their cars. In choosing between two equal products, recognition of the company name and brand names is an important factor. To protect the image values of form and text, the company should apply for trademark ™ protection and should assert its rights by claiming copyright of the product. For similar reasons, Web addresses are important to protect.

7.5 From Need, Want or Wish to Ready Product

7.5.1 Finding Out the Product to Develop

When a need, a want, or a wish for a product has been acknowledged but for which a technical solution does not exist, it can take quite a long time to find a solution on which to base the concept. However, individual and collective creativity, like brain storming, can be used to improve the situation.

Those taking part in such actions first need to be prepared regarding the kind of solution to develop. The acceptance of the mission—the storing of it in the active conscience of the brain—is called "incubation". As soon as the mission is stored, the brain will start to work to find useful solutions on an abstract level. This thinking can be called Brain Aided Design (BAD). For a useful solution to show up—the illumination—it can take a long time to find solutions for the need, want, or wish expressed. To speed up the process, different techniques are used, of which some common ones are shown in Fig. 7.5. When a useful solution has shown up, it can be used as a concept to work on further.

The first method (BAD + PAD) in Fig. 7.5 means individual work starting with Brain Aided Design (BAD) and then making sketches (PAD = Pencil Aided Design). BAD can be done anywhere, for example, sitting in a car, or watching the TV. The thinking is mostly done on an abstract and un-dimensional level. When sketching the solutions have been created, they will initially be on an abstract/principal level. The complex connection between the movements of the fingers holding the pen/pencil, the image feed-back registered by the eyes to the brain, and the processing by the conscious mind picking up solutions from the sub-conscious and un-conscious minds is important for the creative work. The more sketches that

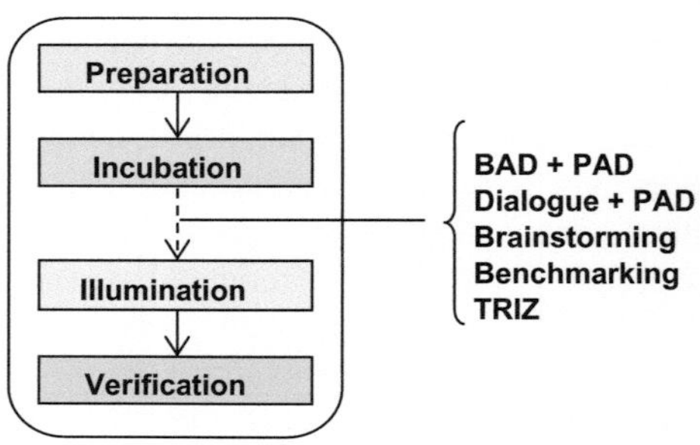

Fig. 7.5 The creative process and some methods to find solutions

are drawn on paper or a white or black board, the more efficient the processing—the creativity—will be. A digital pen connected to a computer has similar advantages as a normal pen/pencil, but the flexibility/availability is lower. Many good ideas have been sketched on the back of an envelope or on a paper napkin at a restaurant!

The computer mouse does not give the same feed-back between the hand and the brain. The eye's focus on the pen/pencil point and the result of the movement of it is different than the movement of the mouse creating a picture on the computer screen. Also, on the computer screen there is, in general, only one concrete picture for the eye to register, which hinders creative thinking. The architect fears the empty paper, is a saying with some relevance.

The second method (Dialogue + PAD) in Fig. 7.5 means to have a dialogue with other people, combined with the making of sketches on a piece of paper or on a white or black board. The coffee break becomes a perfect situation for creative dialogues. Many innovation activities have started with a dialogue over a lunch or a dinner and paper napkins and a pen have supported the creation of big ideas.

Instead of the informal dialogue, a more formal *brainstorming* session can be used. A handful of creative people can sit together saying different words to a secretary who writes them down so that everybody can see them (e.g., on an overhead sheet or a white board). When thinking of each word in connection with the problem to solve, new solutions can pop up. Thus, instead of the eyes registering a sketch, they register a word when brainstorming is used, which gives less good results for technical work but which can be good, for example, when looking for new marketing ideas, or finding possible reasons for different problems.

Basically, to do benchmarking means to investigate solutions already existing to get ideas for what to do. When a problem has to be quickly solved, the easiest, cheapest, and safest way to do this is often to copy an already existing and well-functioning solution. Thus, by benchmarking other solutions and by evaluating different solutions one can often build up useful knowledge on how to solve a 'need' quickly. Knowing what to do also means a good possibility to calculate the development time. For example, Lean Product Development (LPD) has as one corner stone which is the use of benchmarking to get knowledge quickly for bringing new products to the market (e.g., Mehri 2005). The problem with this way of working is that radical new products will seldom occur.

To find the solutions on levels 3–5 in Table 2.3, a method called TRIZ (Altshuller 1980), which is a translation for the Russian "Theory of Inventive Problem Solving", can be used. Using that method, one first generalizes the given problem, instead of searching for a specific solution. The second step consists in finding a general solution to the generalized problem. This is done by identifying contradictions, after which one generally benchmarks each of them using a computer data base with different solutions taken from millions of patents and laws from physics and chemistry. When the contradictions have been solved, one then tries to find solutions to bridge the solutions to get one single product. For example, a lap top is small when folded and large when it is in use, which creates two contradictions to solve, first one by one and then combined with new solutions.

There are two types of basic contradictions (Altshuller 1980):

- *Technical contradictions*: A desirable function A uses a second function B, which has undesirable effects, either causing a third function C which is harmful or harming an existing function D. For example, one can evenly spread light over a large car park by having a tall lamp post. However, this requires a high strength post to hold the large light far above the ground. A (distant light source) needs B (tall strong post) which leads to C (high cost) and D (difficult maintenance).
- *Physical contradictions*: Physical contradictions occur where the two opposing physical states are required. For example, a blacksmith wants the horseshoe to be hot enough so the metal is workable. However, he would also like it to be cool enough to hold (the 'harmful' solution is to use tongs, which are not as easy to use as fingers).

When one or many concepts have been created, it is time to start the engineering design work by making models, followed by prototypes, ending up in production of the products.

[The one who started the work with TRIZ was the Soviet/Russian engineer, Genrich Saulowitsch Altshuller, who indeed had a very interesting life story, which can be found on http://www.aitriz.org/altshuller.htm. Altshuller was born 1927 and died 1998.]

7.5.2 Concept Development

As guides for product development, two types of concept exist: concept product (as a concept car) and product concept (as a car concept). In turn, the product concepts are part of a business concept.

- A *concept product* is created to guide the further technical development or technology development. Industrial technology development is often called R&D—Research & Development, although "Research" leads more to the image of academic research than industrial research, having few demands on following scientific formalities.
- A *product concept* is created to guide the product development of a specific commercial product.

Wish-based PD, and partly also want-based PD starts with developing a concept product (see Fig. 7.6). For the development of a concept product, the market connection is rather weak but it is important for the development of a product concept.

The time it will take to reach a functional level is dependent on many factors. Often disappointments will be felt when a promising solution shows that it will not hold. The right part of Fig. 7.7 shows a principle example of this.

There are many definitions of the terms "models" and "prototypes". Here, a model means a formed solid that demonstrates the function and/or the shape of it.

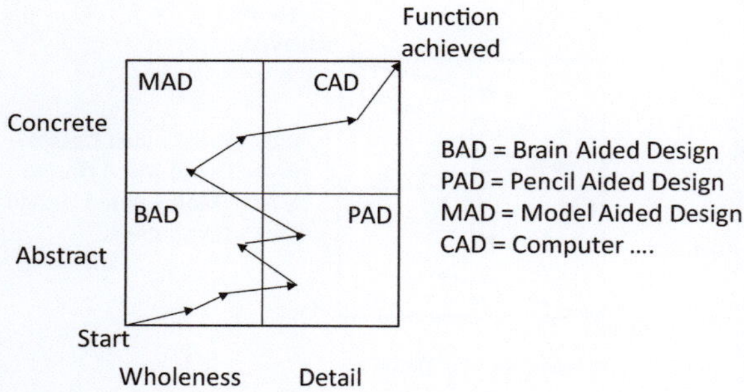

Fig. 7.6 To find a functional solution from a wish, different steps are needed

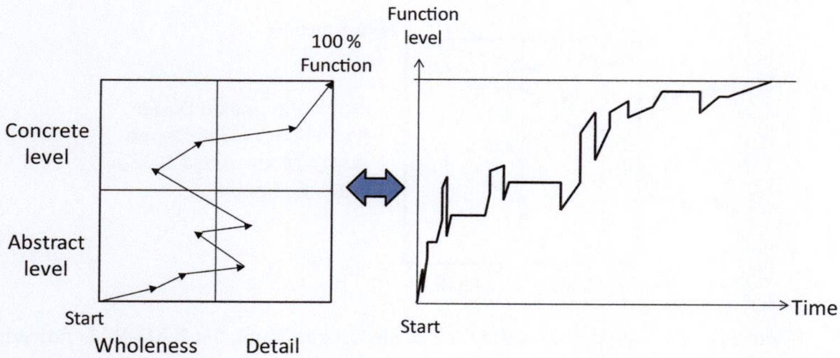

Fig. 7.7 Figure 7.6 transferred to a time-function diagram

The model can be a part of a product concept used in the further development of a new product. When a model exists, a prototype can be developed, which means the development of a production-ready solution. The definition of a prototype can be "An approximation of the product along one or more dimensions of interest (Ulrich and Eppinger 2016)". Here, the development work to make a model is called modelling and the development of a prototype is called prototyping.

When 3D-printing is used for modelling, a CAD-model must be done before the MAD can start. Therefore, in that case, the sequence in Fig. 7.6 must be changed, in principle, to what is shown in Fig. 7.8.

Reversed- or Backward-Engineering can be used as the start of development work in order to have a benchmark of its own solutions. The BAD-PAD-MAD-CAD method has to be changed to what is shown in Fig. 7.9 (see also Appendix C).

Contradictory to what is generally taught—that all demands must be set before the creation of a concept starts—we have found that one should start only with one primary demand and 2–3 secondary demands and then start to create concepts and solutions to satisfy them. When one or more concepts and solutions have been found, more demands are added for each of them. These demands can result in the fact that new solutions must be found. If a solution does not hold in the test and

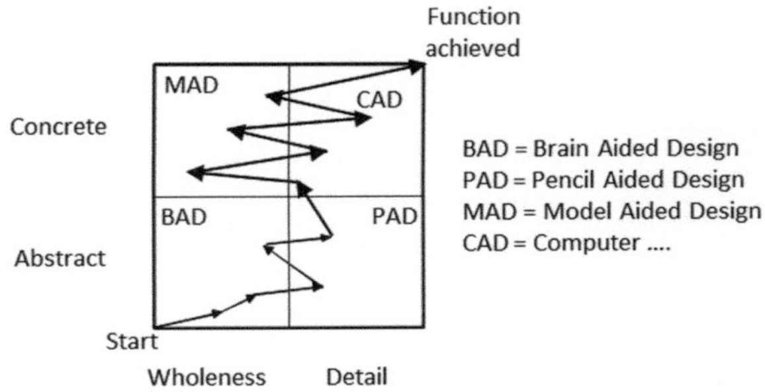

Fig. 7.8 When 3D-printing is used for modelling, the sequence in Fig. 7.5 needs to be changed (Ottosson et al. 2016)

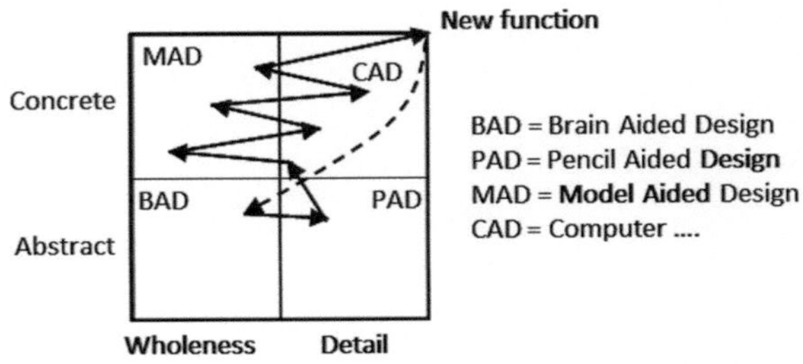

Fig. 7.9 When reversed engineering is the start of the development, the BAD-PAD procedure can be simplified

evaluation, it is stopped from further development and documentation is done of the findings and experiences. As Fig. 7.1 shows, the solution may be picked up again after some time. Using this principle, which is shown in Fig. 7.9, the work can go ahead at a high speed, ending up with a final concept and solution that is well documented. Thus, this final concept is always developed in parallel with the solution time, which is a difference between DPD™ and other PD models for which the concept is static from the time it was once decided upon.

The solutions found after each iteration in Fig. 7.10 that satisfy the demands can be on different novelty levels.

To get information to create a concept, one can use different data collection methods. Figure 7.11 shows some commonly used methods and their reliability. DPD™ suggests that methods should be used that give high reliability, although this means that a classical statistical reliability cannot be achieved. The developers need to make their own experiments and tests to be able to make a fast and cheap development. Smaller companies normally use dialogues, observations and simulations to get material to analyze and make decisions on. Large companies and companies which adhere to the view of first finding out the "voice of the customer" have to use questionnaires and structured interviews to get their statistical material, which means a slow, costly and a rather risky way to work.

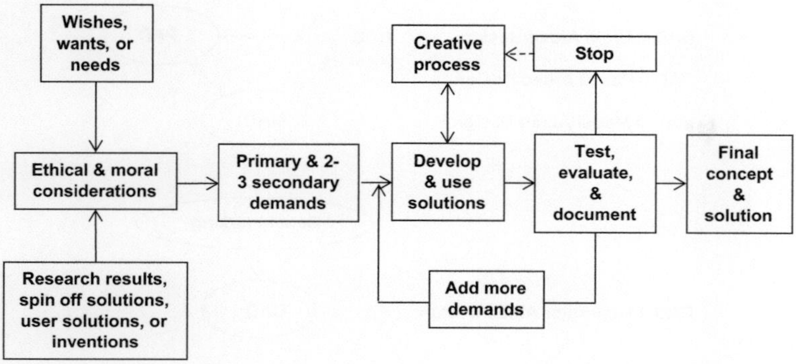

Fig. 7.10 The concept development is an iterative process in DPD™

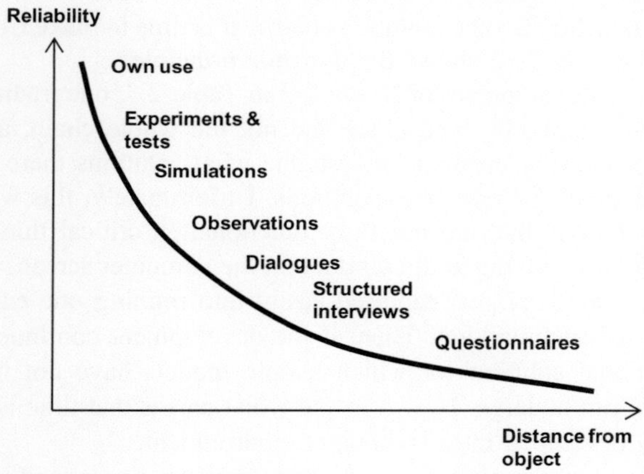

Fig. 7.11 Reliability depending on which investigation method is used

7.5.3 From Concept to Ready Product

When one or more basic concepts, independent of concept type, have been agreed upon as interesting to make further developments, it is time to take on the concrete and detailed development of the product concepts. Initially, and also when problems occur later in the development process, BAD, PAD, MAD (Model Aided Design) and tests of the models appear to be a fast and efficient way of working until detailed engineering design can be done.

To make MAD means to make simple models in as soft materials as possible so as to quickly understand the effects of the solutions with the help of as many of our five "See/Hear/Taste/Touch/Smell" senses as possible. For the shaping and the changing of hardware models in their soft material, in most cases a multifunctional Swiss Army knife is the only tool needed. Techno LEGO® can also be used to find

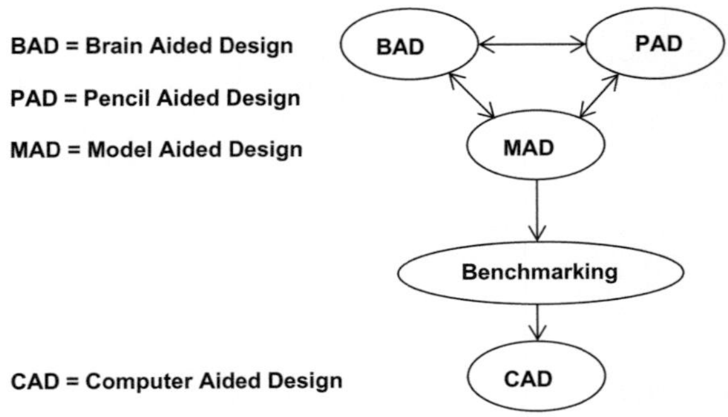

Fig. 7.12 Initially when a concept is developed and later when problems occur, the ways of working shown in the Figure have shown to be efficient (Ottosson 2015)

out the mechanical functions of possible solutions. When solutions have been found, one can benchmark other solutions before it is time to make Computer Aided Design (CAD). Figure 7.12 shows the dynamic principle.

For need-based development of levels 1–3 in Table 2.3, one traditionally strives to make only benchmarking and CAD and not the whole chain, as is shown in Fig. 7.12. *Benchmarking* means to investigate what solutions there exist either to use or to use as input for your own solutions. Unfortunately, this way of working easily brings solutions that are not fully functional as critical thinking seems to disappear when only looking at the product on the computer screen, and thus using only one sense. One also gets captured easily into refining one existing solution when the start is benchmarking. Then, if the development continues too long for only one functional solution for which simple models have not been done and tested, the costs can be large. However, the worst part is that time is lost when the product is only developed in the computer environment.

The ideal situation, therefore, seems to be to develop a new product from scratch to a finished product as shown in Fig. 7.13.

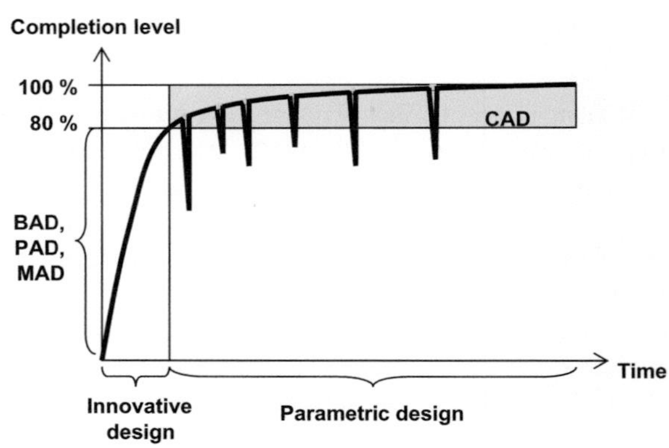

Fig. 7.13 When problems occur in the development of a new product, BAD, PAD, and MAD help to find solutions to continue the CAD work

7.6 Design for X—DfX

Design for "excellence" (also known as DfX or DFX) is a general term used in the engineering world that serves as a placeholder for different design objectives. In the term DfX/DFX, the variable x is interchangeable with one of many values depending on the particular objectives of the venture. Some of the most common substitutes for x include assembly (DfA), manufacturability (DfM), and logistics (DfL).

DfX methodologies address different issues that may occur in one or more phases of the product life cycle:

- Development phase
- Production phase
- Use phase
- Disposal phase

Here, some of all the DfX used in the development phase will shortly be discussed. Traditional Engineering issues as manufacturing and assembly is not the topic of this book, although very important in the development phase as the production phase.

7.6.1 Overall View

Efficient hardware development of different parallel activities can be done as is shown in Fig. 7.14. As DPD™ is a user-centered development model, DfU (Design for Usability) comes first in the development when a functional concept principle has been found. Until the project is finished, checks must be made continuously to ensure that the demands on DfU are not violated. For example, when DfMA (Design for Manufacture and Assembly) shows that a more efficient production will be possible by making changes in the design, DfU must be considered again before settlements can be done on the design. The other objects in the figure used DfXes (or DFXes'). These are 'Ae' for aesthetics, 'Er' for ergonomics, 'C' for Comfort, 'MRS' for maintenance, repair and service, 'L' for Logistics, 'P' for packing, 'R' for Recycling, 'St' for stress, 'En' for environment, and 'Q' for quality. LCA in the Figure stands for Life Cycle Analyzes and FTA for Failure Tree Analyzes.

Initially, when satisfactory values on DfU have been developed, it is time to start to work on more subjective and difficult to measure values that could be called Wellbeing. To that category belongs Design for Aesthetics (DfAe), Design for Ergonomy (DfEr), Design for Maintenance, Repair and Service (DfMRS), and Design for Comfort (DfC). As has been pointed out, subjective values are very important for the success with a new product and the long term economical sustainability of it.

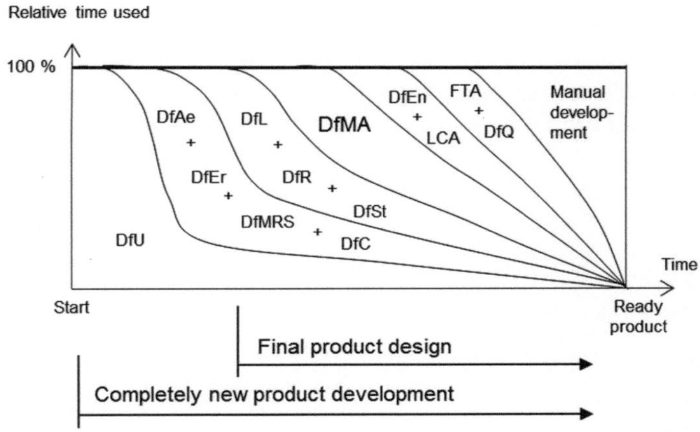

Fig. 7.14 The order in which a new mechanical product is preferably developed (Ottosson et al. 2016)

Thus, as seen in Fig. 7.14, DfU should be present from the start to the end of the development. Therefore, it is of great importance for the product developers to get to "know the user" and the use of the product. She/he also needs to realize that users are not possible to collect in homogenous groups, which is why they require solutions on an individual basis. Age, experience from usage of similar products, or other relevant experience, financial situation, and life situation are just a few of all of the aspects that influence the user of a product. Therefore, the product developer's ability to empathize, participate and understand user situations is critical for the analysis of how new products can support, for example, disabled people and offer adequate usability.

The order in which new DfXes are added after DfU has been satisfied can differ from those in Fig. 7.14, depending on the product and the sub projects in the development work. However, what is important when a new DfX is added and worked on is to control what has so far been achieved and that other values, such as the functional, perceptory, and image values, are not violated.

The time it takes to fulfill the different DfXes is dependent both on the individual team members, how many they are and perhaps most of all, the delay times for information acquisition and decisions. The information acquisition alone can take up to a quarter of the designer's time (Aurisicchio et al. 2006), which is a key reason that the delay time, especially in larger organizations, can be considerable for each team member. The longer the delay time is and the more often it occurs, the longer it will take to reach maturity (see Fig. 7.15).

When the start of a commercial product is a wish, the total development process to get a commercial product can be as is shown in Fig. 7.16.

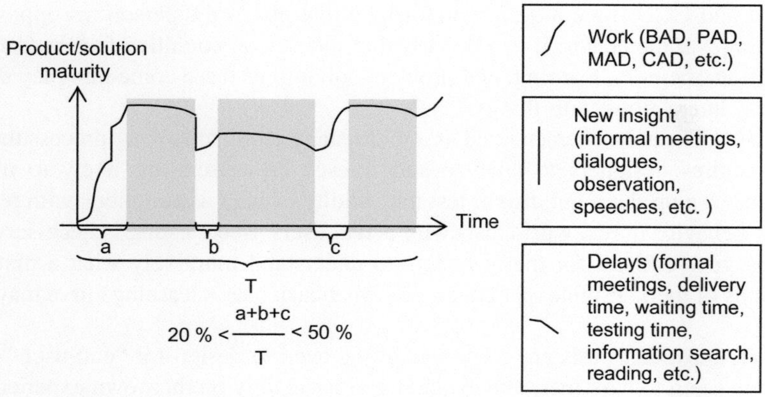

Fig. 7.15 Unfortunately, the efficient work time in a need-based PD project is often in the range 20–50%

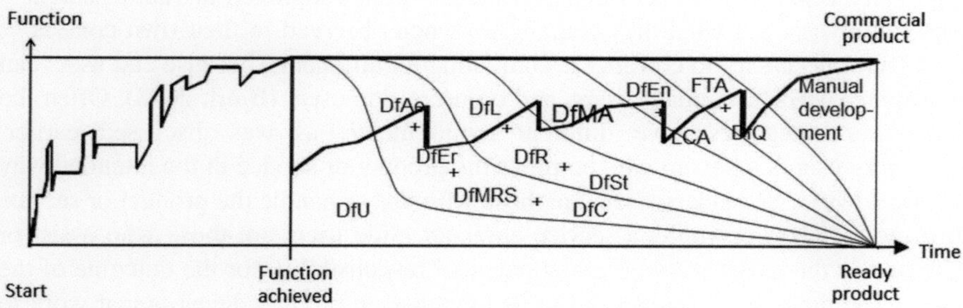

Fig. 7.16 The development of a commercial product starting with a wish means combining Figs. 7.6 and 7.13

7.6.2 Design for Usability—DfU

When useful functional values have been created for the new product, DPD™ is advised to satisfy user/use demands in the development first. Other PD methods mostly place their emphasis on production. Thus, in the detail development, Design for Usability (DfU) should be satisfied first in DPD™, and at all times during the development, the usability should be kept in mind so that it is not neglected or weakened when other demands are satisfied.

Note that in the software industry sector, a DfU-related term is User-centered Design (UCD). UCD has as its aim to ensure that the man-machine interface is user friendly. The main difference from other interface design philosophies is that user-centered design tries to optimize the user interface around how people can, want, or need to work, rather than forcing the users to change how they work to accommodate the system or function.

DfU and UCD share a design philosophy that places the person (as opposed to the 'thing') at the center; it is a process that focuses on cognitive factors (such as, perception, memory, learning, and problem-solving) as these come into play during peoples' interactions with things.

DfU/UCD can be characterized as a multi-stage problem solving process that not only requires designers to analyze and foresee how users are likely to use an interface or a product, but also to test the validity of their assumptions with regards to user behavior in real world tests with actual users. Such testing is necessary as it is often very difficult for the designers to understand intuitively what a first-time user finds of their design experiences, and what each user's learning curve may look like.

While most designers are conscious of the need to design for "end-users", they too often seem to base their understanding of users only on their own experience or on findings from market research. In contrast, user-centered designers also engage with potential users directly to get their user experience (UX), believing that understanding the details of individuals' experience gives greater insight than the aggregated reports of market research, and that what people tell market researchers does not always tell what they actually do when observed in their own context.

A factor necessary to consider is conflicting requirements, because end users can be primary users, secondary users, and co-users/side users (Björk 2003). Often the three user categories have different requirements, as was discussed earlier. A primary user is a person who employs the product or service in the intended way (Hansen 1991). Secondary users are those who use or handle the product or service in some way, for example, a service engineer. Side users are those who assist or take part in the usage without taking personal responsibility for the outcome of the usage. The three user groups need to be identified in product development work to obtain product usability for all categories.

Many standard design models also involve customer and/or user feedback in the latter stages of the product development. However, the user-centered designers need to start engaging with users during the early, formative stages to set the agenda for their projects, rather than waiting until it is too late to make significant changes. Then, during the whole development process to a commercial product, one will never lose sight of the customer's and user's views.

According to ISO (ISO DIS 9241-11), usability is "the effectiveness, efficiency and satisfaction with which specific users can achieve specified/particular goals in particular environments". On a deeper level, as was mentioned in Chap. 3, these three terms have the following meanings:

- *Effectiveness*: Is the product proposed effective for reaching the goal? Is it possible to implement the findings in real user environments? What is required to make that happen (e.g., education needs, training needs, expert needs, tools acquisition, or organizational change)?
- *Efficiency*: Is the product proposed efficient to use? Is it tricky to use? Is it time/resource intensive?

- *Satisfaction*: Will the users find the use of the product more pleasant to use than what they experienced before the implementation? Will the users feel that the outcome is more efficient? Will the use of the new product contribute to a better economical result for the individual or will it reduce failure risks in any aspect?

The development of commercial products that have good usability is a complex task in which 'hard' functional values and 'soft' perceptory, image and emotional values need to be satisfied. Therefore, first satisfying the main/prime functional values and some secondary functional values before the perceptual values are satisfied is important in the product development process. Otherwise there is no reason for developing the products at all.

From when the proper function is secured, the work to establish good perceptual values to a higher level has to be focused on what the products communicate to the users. If the message the products send is falsely interpreted by the users, the risk of misuse increases (Wikström 2002). Also, as the products are normally used in many contexts and by users with cognitive or language difficulties, difficult or abstract product messages should be avoided. In general, if an instruction manual is needed for the correct use of a product, it may be that its perceptual values are not got good enough.

In the product development process, there are often a number of user characteristics that need to be considered (Jordan 1998). Examples are:

- Earlier experience (of similar products)
- Cultural background (language, traditions)
- Special needs or a disability of some kind
- Age, sex and knowledge

Lack of usability can lead to minor frustration as well as life threatening situations for the individual. It has also been documented that many people can have difficulties with everyday items that cause annoyance and defeat their intended purpose (e.g., Norman 1988; Magnusson 2001). In addition, unusable products also cost time and money for the users and customers. For example, difficulties in using computers can cost a company 5–10% of their total working time (Jordan 1998). Low levels of usability could mean that users employ just one product function out of several available. A long time needed to learn how to use the product is also frustrating. High service costs and many customer complaints because of misuse are other signs of low product usability.

The ability to memorize how to use a product is dependent on several personal and environmental circumstances. The user should, for example, obtain memory prompts and cognitive support from the technology. Effort should not be required to handle the technology, and the design should, in the best of worlds, be self-explanatory, thus allowing the user to concentrate on performing the task. An important aspect that aids memory is the ease with which certain features of the construction can be visualized and that they give the right message. For example, a safety belt must make a clicking sound when it is properly adjusted, just as a camera should produce a click when a photo has been taken.

To understand the use of a product, the product developer should, if possible, act as a user and use the product as much as possible during the different development stages. If possible, one or more demanding user(s) should also use the product in its different appearances in the development process and comment on the solutions. This is because it is far better to design with someone, than for someone (Rowland 2004).

The development of products and physical environments has a democratic dimension so that, for example, the result of an architect's work is both new buildings and democratic implications as un-necessary built in barriers will prevent disabled people from being a part of the society on equal conditions with other citizens. Barriers can be stairs that a person in a wheelchair cannot pass or can be a can that a one armed person cannot open without the help of other people. Examples of barriers to information are Web pages that blind people cannot access.

For the development of solutions for a barrier free world, the terms Universal Design (UD) and Inclusive Design are used. Seven UD principles have been proposed to help developers design products and environments for a barrier free world:

1. Equitable use
2. Flexibility in use
3. Simple and intuitive
4. Perceptible information
5. Tolerance for error
6. Low physical effort
7. Size and space for approach and use

We have found that the seven UD principles should preferably also be extended by three other principles:

8. Comfort in use
9. Pleasure/joy in use
10. Intellectually stimulating.

7.6.3 DfAe (Design for Aesthetics)

The noun aesthetics or esthetics is defined as (www.dictionary.com):

- the branch of philosophy dealing with such notions as the beautiful, the ugly, the sublime, or the comic, as applicable to the fine arts, with a view to establishing the meaning and validity of critical judgments concerning works of art, and the principles underlying or justifying such judgments.
- the study of the mind and emotions in relation to the sense of beauty.

The word aesthetic is also an adjective and adverb relating to cosmetology and medicine, such as in aesthetic medicine.

The aesthetics come in when the elements and the entire design is developed to be something more than utilitarian and functional. It is the extra details put into make them visually appealing. Aesthetics are the artistry. In some ways, it is the eye candy whose function is solely to generate a 'wow' or other emotional response. It is an added layer of beauty on top of the usable and functional (http://vanseodesign.com).

A basic understanding of the principles of good design (such as contrast, unity, and balance) is the foundation for creating beautiful solutions.

7.6.4 DfEr (Design for Ergonomics)

The International Ergonomics Association, IEA (iea.cc), defines Ergonomics as: "Ergonomics (and human factors) is the scientific discipline concerned with the understanding of the interactions among humans and other elements of a system, and the profession that applies theoretical principles, data and methods to design in order to optimize human well-being and overall system performance."

Sometimes the acronym, HFE, is used meaning Human Factors/Ergonomics. Dul et al. (2012) pointed out that HFE has great potential to contribute to the design of all kinds of systems with people (work systems, product/service systems).

7.6.5 DfMRS (Design for Maintenance, Repair, and Service)

Within the service sector, the term Design for Service (DfS) is used to develop service items. In this case, the designers' contributions to service development and innovation often emphasize the designers' capability to involve users, acting in and through multidisciplinary teams and using visualization skills in these situations. Most knowledge about development of new services has been treated within a service marketing and management discourse, where emphasis is put on customer integration in the process, and the co-creation of the value proposition—the service (Wetter-Edman 2014).

A wider term is DfMRS, which deals with the problem of making maintenance, repair, and service on products, such as a machine or a car engine. Often the MRS have to be done where there is limited space for the service people to work, where it is cold or warm, where it can be dangerous to work because of radiation, high electrical fields or high voltage, where it means to be in an environment that is dusty or with high sound levels, or where it is windy. A special situation is when space crew is to make MRS on their shuttle in the space.

Generally speaking, there are two types of maintenance in use:

- Preventive maintenance, where equipment is maintained before break down occurs
- Corrective maintenance, where equipment is maintained after break down. This maintenance is mostly used although it is often most expensive because worn equipment can damage other parts and cause multiple damages

Maintenance, repair and service involves fixing any sort of mechanical, plumbing or electrical device, should it become out of order or broken (known as repair, unscheduled, or casualty maintenance) (DLA 2016). It also includes performing routine actions that keep the device in working order (known as scheduled maintenance) or to prevent trouble from arising (preventive maintenance). MRS may be defined as (EFNMS 2016), "All actions which have the objective of retaining or restoring an item in or to a state in which it can perform its required function. The actions include the combination of all technical and corresponding administrative, managerial, and supervision actions."

To carry out DfMRS, much work can be done using a CAD model and its parts by applying rules on minimum space, load, and so forth. By using Virtual Reality (VR) and Augmented Reality (AR), the service people can train for their service actions away from the site of real MRS, such as, for example in space, inside a rector, or in a TV tower in the Alps.

7.6.6 DfC (Design for Comfort)

The comfort in the use of a product or in the contact with it can be on the whole scale from comfort, over discomfort to pain.

Comfort (or being *comfortable*) is a sense of physical or psychological ease, often characterized as a lack of hardship (Wikipedia 2016). Persons who are lacking in comfort are *uncomfortable* or experiencing *discomfort*. For example, a degree of psychological comfort can be achieved by recreating experiences that are associated with pleasant memories as the experience of ice bergs with VR technology when having been burnt. Comfort is a particular concern in health care, as providing comfort to the sick and injured is one goal of healthcare, and can facilitate recovery. Persons who are surrounded with things that provide psychological comfort may be described as being "in their comfort zone". Because of the personal nature of positive associations, psychological comfort is highly subjective (Kolcaba 2003).

Where the term is used to describe the support given to someone who has experienced a tragedy, the word is synonymous with consolation or solace. However, comfort is used much more broadly, as one can provide physical comfort to someone who is not in a position to be uncomfortable. For example, a person might sit in a chair without discomfort, but still find the addition of a pillow to the chair increases their feeling of comfort.

DfC means to design a product to give as much comfort as possible to a person who uses or gets in contact with a product—to increase the wellbeing.

Comfort can be measured on the scale from pain to discomfort to comfort.

7.7 Work Principles

Some rules of thumb are useful in actual PD work in general, and especially in NPD as it is especially complex. Here are some useful rules of thumb, given with explanations.

7.7.1 Make Useful Scenarios

When developing a new product, it is common to describe the demands in technical terms. However, our tests and practical industrial work have shown that it is better to create use scenarios that the product developers can have in their minds during the work.

Thus, for example, if a camera maker that intends to develop a new camera first gives the scenario that the camera shall be used in, for example, a sail boat competition around the world, it is easy for the product developers to imagine what can happen during such a competition. The next scenario, when the first is satisfied, can be that the camera will be used by an alpinist who will climb Mount Everest. The third can be that the camera will be used by a driver in the dessert Dakar rally. The camera should work well in all three environments. The three cases, which also call for empathy on the part of the product developers, will guide each developer in the actual work. (This is a real example from Hasselblad that I gave advice for in their development of a new camera in 2001–2002).

7.7.2 The Framing Principle

When a new product aimed for many user types is to be developed, one either can start to develop it for an average (normal) user or one can start to develop it for the most extreme/difficult users or use, and then check what is needed to make the solution work for a user/use as far away from the first as possible. In the first case, one has to come up with a new solution when a new user appears that differs from the average user. In the second case, having found the two solutions for extreme users/use one can make use of these solutions to easily find solutions for all users between the two extremes. Therefore, in DPD™ one starts to develop products for extreme users/use and situations instead of making single unique solutions.

Note that Universal Design (see Chap. 2) means finding one single solution that fits all users and not one single solution for each group of users. Also, in this case, the framing principle helps to visualize design problems and develop the single UD solution. It has been shown (Björk 2009) that it is much more complicated to make universally designed products than to use the concept of making modules to suit different users. More complicated means also more costs and more time needed to make acceptable solutions.

In practice, framing is done by forming pairs of conflicting demands and finding which intersections are the most difficult, and then finding solutions for the intersections. The example in Fig. 7.17 shows when four extremes are taken into account when developing a new tool, such as a screwdriver. In this example case, D seems to be the most difficult one to satisfy, followed by case C. Next in turn to solve is case A. Case B seems not to be a problem at all to solve, having already solved the three other demands. Having solved the three intersecting demands D, C, and A one easily can solve the problems for E, F, and G. However, if the main domain of users to target is in the marked area where the three users E, F, and G are, the two solutions for cases D and B should be enough to cover the main domain. Probably two standard products will be needed in that case. Sometimes "modules" can be done so that these modules can be combined to give a large number of unique solutions. In this case, the two standard screwdrivers could be one screwdriver with two interchangeable grips.

When many demands are to be met for the new product, the simple representation shown in Fig. 7.17 can be extended with a multi-axes diagram as shown in Fig. 7.18. The smallest values are in the center of the diagram. When different users with their individual profiles have been plotted, two extreme profiles (defined by the inner and outer boundaries) can be plotted as is seen in the Figure. The development can now be done sequentially, solving the two extremes on each axis.

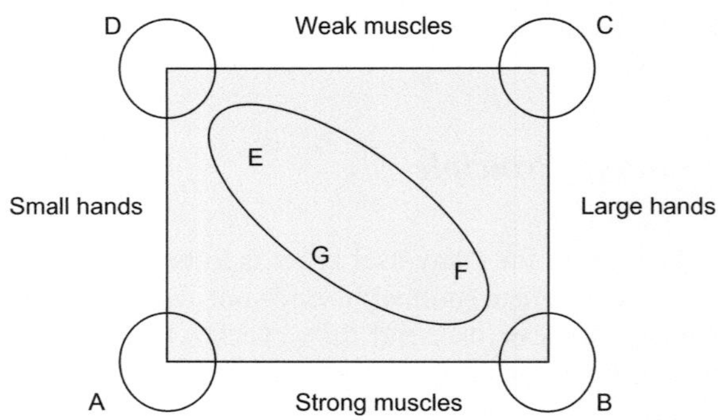

Fig. 7.17 The framing principle means to first find solutions for the most difficult intersections between the extreme pairs. In this case, only two pairs of extremes are shown

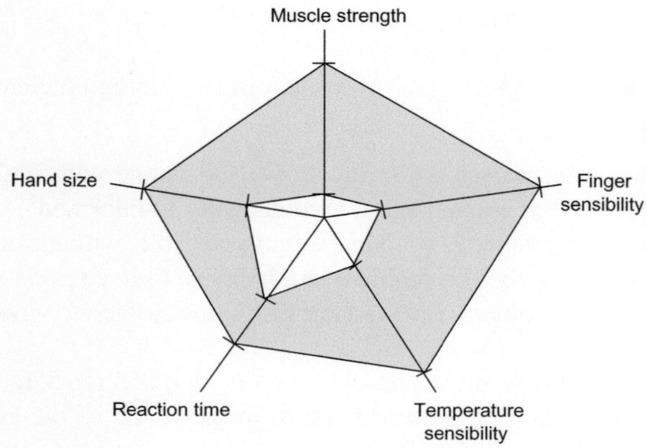

Fig. 7.18 A multi-axes diagram can be used to find extreme values for each demand

7.7.3 Reinventing the Wheel

This is a catch phrase to remind us of the importance of being creative first, before looking at what others have done (Fig. 7.12). If one does not do it in this order, one will be so influenced by what one sees that for a long time, one's own creative ability will be seriously hampered. Sometimes this is called the *Anchoring effect*. Reinventing the wheel is, in general, quickly done but also gives an opportunity to invent a new wheel!

7.7.4 Co-location

"People work together if they sit together", is a saying with some relevance and the importance of co-location cannot be overstressed. Projects failing to yield expected benefits can sometimes be ascribed to this fact. Also, the physical layout of the building is important (Haynes and Price 2004; Olson 2002) as was earlier discussed. "We shape our buildings, and afterwards our buildings shape us" (Upitis 2004).

Teamwork depends heavily on constant communication and in co-located teams. Team members frequently report that some of the best discussions occur spontaneously, based on frequent interactions with co-located workers (Malhotra et al. 2001). Note that we humans have a bandwidth of approximately ten Mbit/s when meeting face to face. All of this, except less than twenty bit/s (about seventeen bit/s), is unconscious communication (Norrestranders 1999) that is impossible to have without meeting face to face.

Co-location of the team yields the following benefits (Holmdahl 2007):

- The team stays focused. No stealing of team members' attention from other groups

- Short communication routs
- Easy to have improvised meetings
- Facilitates the very important overhearing effect yielding efficient spreading of information

If, for instance, the project leader talks over the phone with a client, then the team, by overhearing the conversation, automatically is informed.

This also makes spontaneous problem solving possible, which can happen when one team member hears about a problem and happens to have good suggestions or even solutions to that problem. She/he may also know someone who might have a good solution.

When co-locating the team, the product, or a model, and mockup of the product to be developed should, if possible, be placed in the center of the group. This has many advantages (Holmdahl 2007):

- It reminds the team members of what they are there for, helping them to focus attention on the product.
- It serves as a good visual aid for talks and discussions between team members themselves and between team members and visitors to the team.
- It shows the status of the project if the most recent version is displayed.
- Misunderstandings using product developers with different mother tongue languages will be minimized.

Co-location also means that the project leader can never "hide" in his room, isolated from the team.

If distributed teams are used, it is of uttermost importance that they use the Internet (e.g., Skype) frequently to discuss different matters within the team.

7.7.5 The Principle of Flowing Water

During the technical development, it is essential to always look for emerging main problems and to attack them immediately with as much force as needed. When the main problem has been solved, it is often easy to solve the lesser problems. For smaller problems, the principle should be to go around them and to leave the solution of such problems to a special task (or cleaning up) force. Thereby the progress of the total process is not slowed down by the small problems.

This way of working when the main problem has been solved is called the "Flowing Water Principle". This is because it has similarities to how water flows around obstacles (see Fig. 7.19). The important characteristic is the flexibility of flowing water and its momentum. If the obstacle is massive, water accumulates and eventually finds a weak point and breaks through. In the same way larger, perhaps critical problems are attacked and resolutely solved with the combined force of team members and project resources.

Fig. 7.19 The metaphor of flowing water is used to solve problems without losing momentum (*photo* Stig Ottosson and sketch from Holmdahl 2007)

7.7.6 Switch Between Activities!

The principle of switching between activities is connected to the principles of stepping stones and flowing water. As tempo, initiative and money is lost if people spend their time waiting (e.g., Highsmith 2004) and it is important that they have many activities to switch between. Therefore, if for some reason one cannot continue with what is at hand, one can shift over to the next most important thing to do until it is possible to go back and continue with the first activity/work-task. By successively working in this way, the solutions will be better and better. Also, it has been shown that the creative capability benefits from switching between different activities if they are not too many (maximum six). It has been found that the more experienced and skilled the designers are, the more they iterate between activities (Adams et al. 2003).

Often there is no need to follow a special order when different tasks have to be done to find a complete solution, which is shown in Fig. 7.20, where five tasks are to be completed.

7.7.7 'Lagom' and the Pareto Principle

There is a proverb: "Perfection is the enemy of the good". "Good enough" is the solution to this. This guiding principle can be called the 'Pareto Principle' (e.g., Koch 1998). It says that when working in product development, one should test a solution as soon as it is "good enough". Based on the results, the solution is improved to another "good enough" level after which a new test takes place. After three such cycles, an almost 100% solution has been reached in a very short time. If one instead tries to reach 100% directly, the time it will take to reach that level will be much longer.

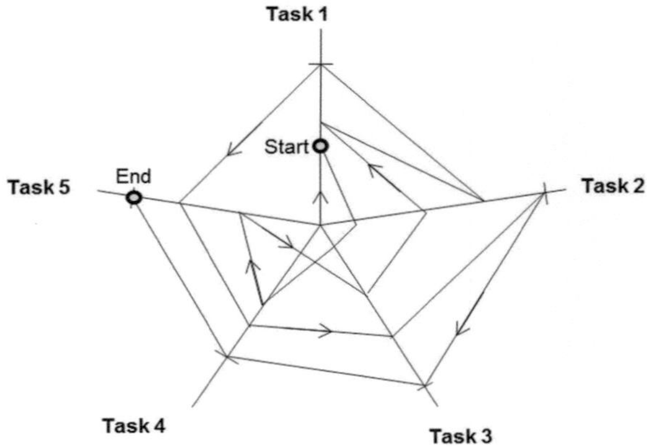

Fig. 7.20 Shift between tasks when good-enough solutions have been reached (the start is in the center and acceptable solutions are at the marks on each axis)

However, in Sweden we have a better term, 'Lagom'. This term is a uniquely Swedish term that does not have an English equivalent. Lagom is "Just enough," "Not too much or too little," "Just right," "Enough to go around," "Fair share." It indicates balance. Lagom does not have the negative connotation of "sufficient" nor does it claim perfection.

[In 1907, Italian economist Vilfredo Pareto observed that 20% of the people owned 80% of the wealth. Often 20% of the stock in a warehouse takes up 80% of the space. 80% of the stock often comes from 20% of the suppliers. Also 80% of the sales will come from 20% of the sales items and 20% of the items will cause 80% of the problems, but another 20% of the items will provide 80% of the production. Experienced project managers have noticed that 20% of the work (the first 10% and the last 10%) consume 80% of the time and resources.

In the late 1940s, quality guru Dr. Joseph M. Juran attributed the 80/20 Rule to Pareto, calling it the Pareto Principle. While it may be misnamed, Pareto Principle or Pareto's Law, as it sometimes is also called, can be a very effective management tool. The 80/20 Rule can be applied to almost anything, from the science of management to the physical world.]

7.7.8 Preliminary Decisions

In traditional management literature, a constant piece of advice is to make decisions as early as possible. By doing so, it is thought the decisions will help to provide orientation for people working in the development process. For management, this also means that it is easy to follow up on decisions that have been put in action. Thus, by taking one step at a time it is believed that the development will be safe and efficient.

However, in reality, the opposite way of acting—making early preliminary decisions and late final decisions—gives a safer and more efficient result (Holmdahl 2007). This does not mean that one should not set deadlines (milestones at specified completion times) as people tend to work harder close to the deadlines than when there is plenty of time to next deadline. This is connected to the principle of "make many small and few large decisions".

One important reason for the principle of making as many late final decisions as possible is that it is impossible to know, in detail, what will happen in the context of the development project. Having made a fixed early decision, therefore, means that the flexibility of the project is taken away as new information cannot be taken into account during the process. Going back on a decision is frustrating and is often seen as bad management. Changing direction more than once often means that the confidence in the project leader deteriorates with every step.

Explaining the reasons for going back on a final decision and motivating the team members for a new decision is a difficult process and is especially difficult if hard facts do not exist as to why a new orientation is needed. When hard facts exist to make a new decision—and not only gut feelings—it often is too late to make the change causing the project to fail.

Thus, in DPD™, as a general rule, one makes final decisions as late as possible. Instead of taking early fixed decisions, one makes preliminary decisions that are easy to change when required without mental blocks. This general rule, of course, must be applied cautiously. It does not mean, for example not ordering models and prototypes on which to make tests, or not hiring the competences necessary to speed up the pace. Needed investments must be taken but, for example, scrap material can often be used for initial tests, which is cheaper and faster than buying new test material.

Taking preliminary decisions means maintaining the flexibility to make changes and take shortcuts when needed without causing mental difficulties. This principle is connected to the next rule—to make many small and few large decisions.

7.7.9 Make Many Small, and Few Large Decisions

In all development, the direction of the work has to be changed now and then. Reasons for changes are, for example, when unforeseen problems and external influences (e.g., new laws, new user patterns, or new technology) occur. A metaphor of that is how to steer a small boat that has an autopilot and which is sensitive to swiftly-flowing water or strong winds. To reach a goal, a new direction must be set every time an external (unwanted) influence occurs as well as when the skipper feels there might be a problem to avoid. Also, whenever short-cuts can be made, they should be tested according to the principles of DPD™.

7.7.10 Design and Verify Concurrently

Not many years ago, engineering design of the part/system/total architecture was followed by the building of prototypes as the only verifying alternative. The prototypes were then tested and test results were analyzed. Often there was no time for redesign in the case of failure during testing, so the design engineer designed the parts sturdy enough to pass the test. This, of course, was a waste of raw material and money.

When CAD was introduced, CAD-drawings and later CAD-models were used for making test specimens. After some time, it was realized that the CAD models could be meshed and used for Finite Element Analyzes (FEA), making the physical testing obsolete.

By using modern software that works in the background of the CAD-program and automatically creates the mesh, the engineer can test and modify the design many times during a single day (Holmdahl 2007). However, in the early development process, MAD and tests often give the fastest development and the best result.

There is an additional meaning to the idea of starting with a coarse concept and then refining it in subsequent steps, sometimes iteratively, and that is that one develops it continuously from the start of the project until it is a finished product. This runs contrary to the established paradigm in engineering design, but is never the less a more efficient strategy for developing products that fit the market situation at product launch. Thus, the mindset should be characterized by preparedness for continuous concept development.

7.7.11 Using the Walls

It is convenient to have large white-boards on the walls and video projectors for projecting, for instance, CAD-models on the white-board. Then the team can draw alternative concept solutions on the white-board on top of the projected image.

Highlighting important information is valuable as, due to the excess of information we get via the Internet and Intranet, we sometimes have difficulties focusing on important issues. Therefore, highlighting important information in other ways becomes important.

While modern Intranet systems make continuous tracking possible, "writing on the wall" (see Fig. 7.21) has shown to be a good complement as not everybody reads messages on the Internet or Intranet and important information, not clearly flagged as such, can be dumped as spam information. Whiteboards are also useful for brainstorming and other creative meetings. (The term War room comes from the strategy room PM Churchill used during WWII in which maps of Europe hung on the walls.)

Fig. 7.21 An example of a "war room" (www.workdesign.com)

Thus, an excellent way to use empty space on the walls is to plot out time plans in as large a format as possible and then hang them on the wall next to where the team is located. Then they are always visible and will serve as a constant reminder (Smith and Reinertsen 1995). Changes to the time plan can be written directly on the plot, for example, with a red marker, making them very visible to the team. Other print outs to hang on the wall are important information for extra attention. Such information can be:

- Descriptions of the user of the product and pictures of its use
- Pictures showing the styling and environment where the product is used
- Plot out drawings such as assemblies, sections, and matting surfaces hanging on the wall together with marked conflict areas, unsolved problems, sketches, or alternative concepts

Whenever a team member leans back or raises her/his eyes, their sight will fall on the wall, meaning that the brain fills up with visual information that feeds the creative process of the subconscious mind.

7.7.12 Notebooks

As described, every team member and the project leader should be encouraged to make notes in her/his own notebook from meetings, when ideas or problems have occurred. Unfortunately, this good piece of advice becomes more and more difficult to follow when we use smartphones for more and more topics. However, looking back in the notebook now and then allows one to reflect on the work done and why certain decisions were taken. One's own wording is more useful than notes from a meeting secretary!

Thus, the records in the notebooks represent a large amount of design information and design related information such as:

- Fundamental design knowledge
- A rich source of design information and rationale that supports decision making
- Information for design audit purposes and a legal record for accountability and Intellectual Property issues
- The results of analysis and modeling including failure as well as success
- Informal information regarding suppliers and customers, including contact details and performance assessments
- The outcome of discussions with experts and colleagues

Clearly, the ability to access this wealth of potentially valuable information is highly desirable, which is why the scanned books can be used and stored in the W-Briefs, as discussed in Chap. 4. Doing so means:

- A more complete understanding of previous design issues and how they were resolved
- Information describing lessons learned and important background or contextual information
- A single accessible location for ideas, sketches and notes relating to a given project or design, improving the 'collective memory' of an organization
- Better support for concurrent and distributed design activities

Over the past decades a number of technologies for recording and managing written notes have been developed. They can be broadly split into:

- Enhanced or augmented paper-based logbooks
- Computers with pen-based input
- Keyboard- or web-based logbooks

None of these technologies can easily support the full range of information recorded in paper based logbooks and are particularly poor at supporting browsing (McAlpine et al. 2008). In addition, there are still some advantages with paper, such as simultaneous reading and writing and reminding through physical presence. Augmented paper-based systems are better in this respect, although they have their disadvantages, the primary one being synchronizing paper and electronic instances.

Thus, in general, electronic technologies do not provide comprehensive support for the full range of essential engineering tasks, such as freeform sketching and annotating external documents (e.g., CAD drawings). Other more routine but important practical requirements, such as portability, survivability and start-up time may also prove to be fundamental barriers to the uptake of an electronic logbook.

Therefore, the use of notebooks is still recommended. Notebooks also have a legal implication in case patentable inventions have been produced in the work. If many apply for the same solution at the same time, the notebook can be the evidence needed to save the patent rights. Because of that, it is important to also note dates and not to use a pencil when making the notes. This as information can easily be changed afterwards. Even when electronic storing is used, paper and physical copies should be saved in case the digital media breaks down or become inaccessible in the future (e.g., from magnetic memories to CDs).

If accidents happen with the products while in use, the notebooks can be a valuable source of information in court trials to find out whether the fault was caused by misuse or product deficiencies. Going back to the notebooks can, in such cases, provides evidence as to why a certain solution was chosen.

7.7.13 Weekly Reports

At the end of every week, each team member should briefly report that week's work results and time, money and other resources spent, together with a brief plan for the coming week. The value of plans is to create preparedness for future actions (Cunha and Cunha 2002).

According to the principles of the planetary organization, the weekly short reports should be sent simultaneously to the project or sub-project leader, to the other team members in the group, and to other people who need fast and direct information.

7.7.14 Information Systems

It is well-known that knowledge gained in one development project is difficult to transfer to another and to coming projects. So far, CAD drawings and PDM files are the main vessels for transferring information. However, they only inform about the final solution and very little about how the solutions came about. Information about the trials and errors needed to find workable solutions are left out, except for information of who has made revisions and when that was done.

To improve the efficiency of transferring information within the company, some larger companies have invested in building up expert systems tapping off knowledge from experts into data banks. However, it has been found that young and rather inexperienced developers regard such information as old and difficult to use in their specific cases. Therefore, they prefer to seek information from dialogues with more experienced colleagues instead of consulting the computerized expert system built up at great costs. Unfortunately, when experts retire or change jobs, it might be difficult to make contact with them when that is needed.

To enhance the efficiency in projects in a company, some measures can be taken with regards to knowledge gained. One is to implement the planetary organization with organized support in the form of using maestros (comets) moving around having dialogues with the team members. Another is to encourage each team member to make notes in their notebooks and see to it that the project assistant or staff fills in logbooks from the collective work done in the project and its sub-projects. A third possibility is to develop systems with knowable people to contact.

However, to reach the level when individual and collective knowledge in an organization is used on a broader scale, K-Briefs (Knowledge Briefs) (Kennedy 2003) and Web-based W-Briefs (Wisdom Briefs) can be used (see Chap. 4). These briefs are filed by the project leader(s) or their assistants when reflected knowledge is at hand and should contain successful findings as well as other findings that can make new projects more successful.

7.8 Verification

As has been pointed out, it is extremely important to verify the different types of solutions that have been created. This is because it becomes more and more difficult and costly to make necessary changes when the sequences are longer between the tests and simulations. It has also been pointed out that verification with only one human sense is less reliable than using many of our five "See/Hear/Taste/Touch/Smell" senses.

Therefore, it is better to have a model to touch, even a very simple one, than just a sketch, a drawing, a picture or an animation to look at. This since the eye often deceives our judgment even when the presentation is on paper in scale 1:1. This counts also for virtual reality (VR) although this has the advantage of involving more than the one sense.

Thus, it is important to produce models and prototypes that the product developers and test people can hold in their hands or touch. If models are impossible to make from a practical point of view in scale 1:1, we have found the old recommendations of scales 1:2, 1:4, 1:10 when making paper drawings in ancient time to be acceptable.

7.8.1 Modelling

To be able to better understand a product idea, models are made either as a physical model, which is sometimes called Physical Mock-Up (PMU), or as a virtual model, which is called a Digital Mock-Up (DMU). Another possibility is 3D-printing of models and prototypes.

If 3D-printing is not used, early in the development of smaller products or scaled down versions of larger products soft material should preferably be used to make physical functional models. The strength of rough modeling materials is also that they encourage playing with ideas, possibilities and potential at low cost. This is because simple material is quick and easy to change. From our experience, clay, paper, wood, and so forth should therefore be used before using harder material. Quote: “When a model starts to harden up, so also does the thinking.” (Schrage 2000, p. 79).

Our experience has shown that the more people involved in the development process, the more important it is that they work with physical models and prototypes. This is due to the fact that models and prototypes are produced to answer questions and to give impulses for development that cannot be gained from virtual models. It is also important to bear in mind that the making of models and simulations has an internal political dimension, which is that managing modeling and simulations is about managing power and influence.

The productivity of prototype-driven design measured in user satisfaction per man-hour has been shown to be ‘superior’ (Schrage 2000, p. 73). One reason is that the product development teams and the test users get the same mental picture to work from when they are all able to touch the model and later also the physical prototypes. For design studies, especially of large objects, digital mock-ups serve the same purpose. Models are thus important visualization tools for developers, managers, users, and customers. Models help us to get a better understanding of ourselves and our priorities and help us to avoid mistakes and misunderstandings.

7.8.2 Prototyping

After the model has been refined to a satisfactory level, it is time to make prototypes and thus, start to close the gap between an idea and a production-ready product. Then it is usually the time to make the development as Computer Aided Engineering (CAE).

As a rule, the more simulations, prototypes and prototyping cycles used per unit of time, the more useful and technically perfect the final product will be. Note, however, that frequent prototyping easily leads to adding an increasing number of features to make the prototype even better so that the product may finally have more features than the user actually needs or wants to pay for.

To cut down development costs and to shorten 'Time to Market', many attempts have been made to reduce the number of prototypes in the development process. Such business cultures are sometimes called specification cultures ('spec-cultures'), which are the opposite of prototype-driven cultures, of which DPD™ is an example. According to our observations, it is not unusual for companies to even spend thousands of hours developing detailed specifications that are invalidated by the initial prototype or the initial market confrontation. 'Spec-driven cultures' draw heavily from market research data before concepts are moved into the prototyping cycle. In prototyping cultures, prototypes are typically used to elicit market feedback well before final versions of the product are tested.

As soon as possible, the functional products should be tested by people other than the product developers who have designed the product. Almost always, it turns out that the result of the use by "outsider" users will be different from that of "insiders". Things will break down and bugs in the software will show up that did not show up in the internal testing. The product and its software will also be used in other ways than thought of or imagined. This eventually will call for a reengineering of the product. Sometimes the product developers will realize that they have developed unnecessary functionality that can be taken away.

7.8.3 Simulation

Simulation is the imitation of some real thing, state of affairs, or process. The act of simulating something generally entails representing certain key characteristics or behaviors of a selected physical or abstract system. Simulation is used in many contexts, including the modeling of natural systems or human systems in order to gain insight into their functioning. Other contexts include simulation of technology for performance optimization, safety engineering, testing, training and education. Simulation can be used to show the eventual real effects of alternative conditions and courses of action.

Computer simulation is used to model a real-life or hypothetical situation on a computer so that it can be studied to see how the system works for pre-defined situations. By changing variables, predictions may be made about the behavior of the system. In PD, computer simulations can be used to perform validation tasks like these:

- Study the performance of the designs for excessive deflections and stresses under dynamic loads.
- Perform nonlinear analysis—including impact—on plastics, rubbers, polymers, and foam.
- Conduct contact analysis coupled with nonlinear materials.
- Evaluate the behavior of composite materials.
- Simulate forced vibrations. Predict and control vibrations or dynamic responses in the products.

- Assembly simulation. Study the interactions of assembly components on-screen.
- Mechanism simulation. Check, for example, for colliding parts.
- Product failure prediction. Predict structural failure thresholds due to yielding, overheating, buckling, and fatigue.
- Compare and optimize alternatives. Determine the best design option by comparing strengths, life, cost, and weight.

The number of computer simulation possibilities increases rapidly, meaning that better products can be the result when it comes to satisfying functional values. However, key issues in simulation include acquisition of valid source information about the referent, selection of key characteristics and behaviors, the use of simplifying approximations and assumptions within the simulation, and fidelity and validity of the simulation outcomes. Therefore, real verifying tests are important to do to cover unforeseen situations.

7.8.4 Rapid Prototyping

Rapid Prototyping, Freeform Fabrication (FFF), or 3D-printing/Additive Manufacturing, are names given to a number of related technologies that are used to fabricate physical objects directly from CAD data sources. These methods are unique in that they add and bond materials in layers to form objects. Such FFF systems are also known as Solid Freeform Fabrication (SFF) and/or layered manufacturing. They offer advantages in many applications compared to classical subtractive fabrication methods such as forging, milling or turning:

- Objects can be formed with any geometric complexity or intricacy without the need for elaborate machine setup or final assembly.
- Objects can be made from multiple materials, or as composites, and can even be varied in a controlled fashion at any location in an object.
- Solid freeform fabrication systems reduce the construction of complex objects to a manageable, straightforward, and relatively fast process.
- Simultaneous production and assembly means that products impossible to manufacture otherwise can be produced.

These properties have resulted in their wide use as a way to reduce Time to Market by getting product models faster than traditional machining of the models and prototypes. Thus, they are used by engineers to better understand and communicate their product designs. Surgeons, architects, artists and individuals from many other disciplines routinely use the technology.

Free Form Fabrication (FFF)/3D printing/Additive Manufacturing (AD) technology is used for Direct Manufacturing of fully dense parts from metal powder. The technology is, in that case, based on Electron Beam Melting (EBM) and parts are manufactured by melting metal powder and building up the parts layer-by-layer.

Thus, with Rapid Prototyping the fabrication of custom objects can be made when 3D computer files exist of the object. The basic operation of any 3D printing system consists of slicing a 3D computer model into thin cross sections, translating the result into 2D position information, and using this data to control the placement of solid material as plastics, metals, and cement but also as human skin and other biological material. This process is repeated for each cross section and the object is built up one layer at a time.

Additive Manufacturing (AM) has historically been associated with manufacturing environments, where it is used for the rapid production of visual models, low-run tooling, and functional objects. However, the impact of 3D printing goes far beyond these applications. Therefore, the technology is in increased use in fields as diverse as biomedical engineering, dentistry, electronics, aerospace, architecture, and archeology.

Appendix C shows an example of 3D-printing and some benefits of using that method for creative modelling.

7.8.5 Scanning

When a physical object exists that is going to be further developed, manufactured in more than one copy, or simply better documented, it is necessary to get the information of the surfaces digitalized. This can be done by scanning the object and sending the measured points to the computer. Such scanning can be made optically (non-contact scanning) or by having a needle that touches the surfaces of the object (contact scanning). Contact scanners, 3D laser and fringe projection based scanners are precise to around 0.1 mm.

Although it may seem simple to have all the equipment and software needed to scan an object, there is a complication, and that is that CAD systems use vector files while scanners produce raster files. Raster files are fundamentally different from vector files. If one draws a circle with the CAD software, it is stored as a circle. The software knows the radius, line thickness, line type and center point. The circle is 'intelligent' because any part of the circle 'knows' that it is part of the circle and 'knows' what the rest of the circle looks like. The raster file created when an object in three dimensions (3D) or a drawing in two dimensions (2D) is scanned, the scan consists of many individual dots or pixels. The scanned image therefore is 'dumb' because the pixels form a shape but each pixel does not 'know' that it is part of the shape. Therefore, transformation work from the raster information to the vector world is needed to make it useful for further CAD work. Such work is difficult to do automatically so many engineering hours often are needed to go from a scanned object to a 3D file of the object.

7.8.6 Virtual Reality

Virtual Reality (VR) means that an individual encounters 'synthesized' experiences created by computers and performed in such a way that the experiences, to some degree, are experienced as real by the user. The individual will, through uncontrolled and controlled reactions in real time, influence the computer generated VR. Virtual refers to its computer-generated existence; some prefer the term "cyber" to reinforce the point. Reality is a more controversial term. Realism debates the whirl around what levels of realistic detail are needed and affordable. Practitioners can choose types and amounts of reality varying from objective to novel and from specific to variable, or nonspecific.

VR has five main components which are variable according to the instructional context requirements:

- dimensionality,
- motion or animation,
- interaction,
- viewpoint or frame of reference, and
- immersion, or embodiment, through enhanced multi-sensory experiences.

As VR uses 2D technology and as CAD files normally are 3D, some work is needed to go from CAD files to VR files and vice versa. Input for a VR presentation is often CAD files but going direct from BAD (Brain Aided Design), PAD (Pencil Aided Design), and MAD (Model Aided Design) creating VR files is possible. The presentation of the VR can be done via HDM (Head Mounted Displays), VR goggles, Cave (e.g., a large cube in which the user is enclosed getting the VR pictures on its walls), as well as combinations of these.

By combining a Content Management System (CMS) for the handling of Web pages and a VR file transformed as a compressed VR file, the VR application can be used as an ordinary Web application. This, in turn, means that more people can be involved in the decision-making process of a new product design, a new production line, a new city plan or simply the purchase of a product. If the possibilities to chat over the Web page are used, a fast market investigation at virtually no costs can be carried out instead, for example, of the heavy to use and costly method of Quality Function Deployment (QFD) and other methods for collecting and analyzing customer opinions about a product.

The sequence from having an ordinary VR file to the publishing of it on a web page is:

- Compress the VR file,
- Import the VR file to a CMS with a web module,
- Publish the VR file on a web page.

The step from having engineers and designers as the only users of a VR application to letting anyone, without geographical restraints to use it opens up new dimensions in decision making, not only in industry but also in society as a whole, where Web based VR, for example, in urban planning could help deepen democratic decision-making. What is needed for that step is to simply develop and add a VR-module to an ordinary content management system.

7.8.7 Augmented Reality

When (real) reality is combined with Virtual Reality, we get what is called Augmented Reality (AR). The determination of Augmented Reality compared to Virtual Reality (VR) is achieved by the direct integration of the real surroundings, for example, in the form of a live-video stream. Thereby considerable parts of the former digitalization effort can be cut off using AR-technology as they will be replaced by digital photos or videos.

With AR-systems, visual 3D-information can be integrated seamlessly into the real surrounding. Powerful solutions allow the realization of the AR-visualization in real-time, whereas the field of view of the user will be extended by context-specific virtual information. AR and its applications can be used to support users at their work-related tasks, minimize digitalization efforts or enable innovative ways to visualize products. With AR, virtual information content will be communicated in a more user-friendly and effective way and in direct context with the real environment.

From the software engineering side, an AR-system consists of three basic components:

- Tracking component, for example optical camera tracking,
- Database system with scene generator,
- Visualization component, for example, software-rendering.

The ability to see the superimposed data and see the real world at the same time allows workers to be head-up and hands-free as they do their jobs by receiving just the information they need, where and as they need it, with complete situational awareness. Workers are able to stay focused on the task rather than having to look away at a handheld device or by walking back and forth to a computer terminal.

To achieve this, the HMD (Head Mounted Display) or VR goggles optically guide an image directly to the user's eye. The specially-coated ocular piece is optimized to allow the image to be reflected into the user's eye, while simultaneously allowing the user to continuously see the outside world unhindered. Users can adjust the optical focus of the image, placing it precisely at her/his working distance. The result is striking clarity of the combined image data and the real-world used, for example, to train service engineers around the world, without bringing them to a training camp.

7.8.8 Pugh Matrix Analyses

When there are many criteria with different levels of importance to take into account to rank different alternatives in life, a *Pugh matrix* can be useful. An example is to evaluate various technical solutions in a PD project (Pugh 1991) as well as in other situations where there are many alternatives, such as to evaluate the geographical market to introduce an innovation in, which includes financing an alternative to concentrate on the development of an innovation, where to locate the business headquarters and production site, who of different people to engage for a critical task, and so forth.

The use of the Pugh matrix is best demonstrated with an example comparing five solutions A, B, C, D, and E to find out which is the best to go further with. To do that, we first decide on criteria for the evaluation of the five solutions. We can start, for example, with four such criteria (see Table 7.1). If we have a solution E that we can use as a solution to compare against for the different solutions, we can call that solution the "Baseline". For each of the alternatives, we then give the numbers 0, −1 or +1 where 0 means an equally good solution to E, −1 a less good solution and +1 a better solution. Having done that, we can sum each column and find out that alternative D is the best of the five alternatives in the example. Solutions A and C are equally good. Solution E is, in this case, better than solution B.

However, we might think that the judgement is too rough and that we want to give a weight factor for each criterion to multiply the factors with. Table 7.2 shows such an example. Now it turns out that solutions C and D are the best solutions with

Table 7.1 Simple Pugh matrix used to find out which solution is the best

Criteria	E-base	A	B	C	D
1	0	1	−1	0	1
2	0	0	−1	0	0
3	0	1	1	1	0
4	0	−1	0	0	1
Sum		1	−2	1	3

Table 7.2 The Pugh matrix in Table 7.1 extended with weight factors to get a more fine-tuned evaluation result

Criteria	Baseline	Weight	A_B	A_W	B_B	B_W	C_B	C_W	D_B	D_W
1	0	2	1	2	−1	−2	0	0	1	2
2	0	4	0	0	−1	−4	0	0	1	0
3	0	7	1	7	1	7	1	7	0	0
4	0	5	−1	−5	0	0	0	0	1	5
Sum			1	2	−2	1	1	7	3	7

the same score, so we need to discuss the two solutions more carefully and maybe add more criteria.

(The Pugh Matrix was originally developed by Stuart Pugh who was a professor and head of the design division at the University of Strathclyde in Glasgow. The Pugh matrix is also called different names, such as, Pugh analysis, decision matrix method, decision matrix, decision grid, selection grid, selection matrix, problem matrix, problem selection matrix, problem selection grid, solution matrix, criteria rating form, criteria-based matrix, and opportunity analysis.)

7.9 Manuals

For most products, a good manual is needed so that the product will be used as intended and also protect the user and the environment from effects of the wrong use of the products. Note that the manuals produced are also legal documents that can be used as evidence if a severe accident should occur. Therefore, primarily for these two reasons, it is important to ensure that the manuals are written properly.

The production of good manuals is tight and concise work. For written texts, the following example can be used (Alströmer 1998):

- Writing of the first edition: 6–10 h/page,
- First review by two reviewers: 0.5–1 h/page and reviewer,
- First to fourth correction: 0.5–1 h/page,
- Second to fifth review: 0.5–1 h/page and reviewer,
- Final review by the author and 3–5 reviewer: 0.5–1 h/page and person.

In addition to all this work, there have to be meetings and dialogues on what to write and how to present the message. It is also best to bring some users into test the manuals and observe how they treat the message.

To supplement the text, or the other way around, it is advisable to make illustrations or take photos/videos of the use of the product. On the Web page, the photos and videos can be shown together with the manuals. The big advantage with good illustrations/pictures is that they are language independent and give the user immediate information that is difficult to explain in words. The old saying that a picture tells more than 1000 words is worth keeping in mind when writing manuals!

For modular based systems, such as a car, the manuals should be only what the customer will get and not everything (all modules) that exist for that car model. With print-on-demand that is easily done.

Chapter 8
Marketing New Products

8.1 Introduction

Marketing is the means by which companies and organizations employ a range of strategies to create interest in their products and services and to help the sales people to sell the products. The start of the development of a market strategy is to identify possible target customer bases and then follow this by concerted campaigns in which the aim is to fulfil the customers' needs, wants or wishes and to build long term loyalty.

It should be mentioned here that marketing does not need to have a commercial aim. Non-profit organizations, such as Greenpeace, Rotary International, religious movements, and gay movements pursue three aims of marketing:

1. display the product,
2. create interest,
3. persuade without normally dealing with price (except possibly for a yearly membership fee), letting the "customers" pay whatever they want to support the organization and its mission.

Such pressure groups can be significant to the marketing of commercial products, since they are often trendsetters. Also the financing of new products and businesses via the contributions of many supporters found through social media—so called crowd funding/sourcing—is an example of this. (The crowdfunding model is based on three types of actors: the project initiator who proposes the idea and/or project to be funded, individuals or groups who support the idea, and a moderating organization (the "platform") that brings the parties together to launch the idea.)

In the case of Greenpeace, for example, its activities have contributed to increased environmental awareness. Rotary International's vision of exterminating polio in the world has, in a similar way, led to new ways of not only vaccinating children in large numbers in poor countries, but everything around it, such as

S. Ottosson, *Developing and Managing Innovation in a Fast Changing and Complex World*, https://doi.org/10.1007/978-3-319-94045-8_8

logistics, and fund raising (www.rotary.org/en/polio-pledges-2017-convention). Entrepreneurs generally need to identify trends and trendsetters at an early stage, as they provide conditions for new products and innovative businesses.

For new products and new services, there are some differences from the marketing of "old" products and services. However, the tools used are the same in both cases. This chapter has as mission to give some important insights related to innovation in the ever-growing topic of marketing.

8.2 Market History

The word "market" is derived from the Latin "mercatus", which means "trade". In the classical era, there were marketplaces where suppliers displayed their goods, created interest in the goods, and persuaded passers-by to buy the goods. This last step was then—as is today in a more developed form—crucial for the outcome.

However, the decision a customer makes to buy a product or not is dependent on many things, including which buyer category the customer belongs to, the customer's present and/or future need of the product, the solution(s), the customer's emotional value of the product, the total price of the product during the time it is in the customer's possession and use, the customer's financial situation, the estimated further sales price, and the positive or negative value of it at the time of its disposal.

Today the marketplace is no longer as defined and clear-cut as it was in the classical era. For example, the Internet has become the leading global marketplace where the products are displayed and explained, where interest is created for the products, and where buyers and sellers agree to transactions without meetings and with no persuasion needed in most cases—at least when customers are offered the possibility of Web shopping. In the classical era at well as now, the message—the content—was/is important as well as persuasion. At an early date, the art of persuasion—rhetoric—became an important skill and the Greek philosopher Aristotle (384 BC–322 BC) is reported to have said that: "Rhetoric is the art of finding that which best serves to persuade—whatever the issue".

Rhetoric in the form of instantly speaking persuasively is today used, for instance, by sales people, teachers, politicians, club leaders, attorneys, and pastors, while market rhetoric is often developed by marketers. Normally the marketers have plenty of time to develop a selling message as well as other supporting activities, such as pricing and displaying the products in a way that will catch the interest of potential buyers, members, or voters. Thus, speaking rhetoric is an instantaneous face-to-face (F2F) activity, while market rhetoric is a non-face-to-face business.

8.3 Aims of Marketing

The basic aim of modern marketing is to:

- create a general positive interest for the products and services, and the organization behind it,
- develop a trustworthy and sticky message,
- make potential customers want or wish to buy the products/services,
- make it easier for the potential customers to place an order,
- make it easier for the sales people to persuade people to buy the products/services.

Marketing is done, to some extent, by all employees in the company. The switchboard operator can (when such an old device is used) influence, in a positive way, the potential buyer's attitude towards the company if the caller is received swiftly, pleasantly and efficiently. A negative reception can have the opposite effect, but even a negative switchboard operator is probably better than an automatic answering system, which unfortunately is rather common for larger organizations, asking the caller to press buttons step by step.

Common experiences and research have shown that engineers in companies selling industrial products often have closer and more long-term pre-purchase relations with customers than sales personnel and marketers. Experience, however, tells us that the same engineers can destroy business opportunities by discussing prices with customers. Thus, it is important to get engineers to let the sales people take care of pricing when a sales force has been set up.

To fulfill these aims, many activities are needed. Figure 8.1 shows an example of a mature organization taking care of different activities within the field of marketing.

During the innovation process—when marketing a new product—the entrepreneur has to initially cover all areas successively, bringing in people who take care of the different activities as the business grows.

8.4 Customer Considerations

8.4.1 Customer Loyalty

As discussed earlier, a customer may be both a consumer and user. The customer pays for the product or is authorized to sign a bill to be paid. The user is someone who uses the product once, but also many times or over and over. To be a consumer means to consume the product, which, in turn, means that the time during which the consumer utilizes the product is often short (e.g., one day for a medicine tablet with a depot effect).

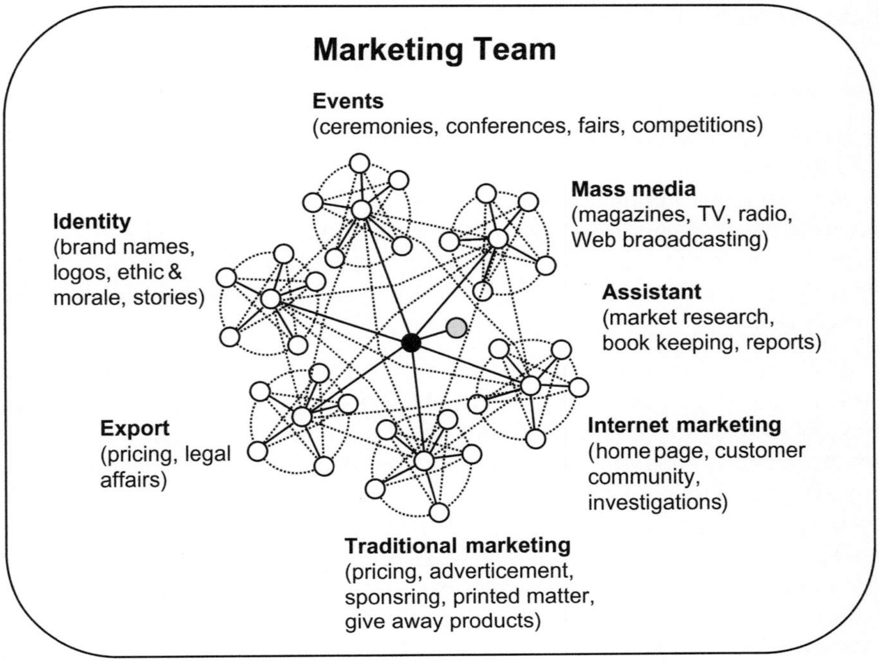

Fig. 8.1 In a mature market organization many activities have to be covered

The terminology covering everyday usage is often used carelessly. For example, can the users of a microwave oven or a car be consumers of such products? There is no doubt that they are consumers of the energy required to operate the products, but the microwave oven is not used up, so a more natural term should be users of the microwave oven instead of consumers. If a person rents a fancy car for a week-end, just to feel how it is to drive such a car, she/he in that case is a consumer. If she/he buys the car after the test, she/he will be a customer as well as a user, independent of whether she/he rents the car, takes a loan to pay for it or pays with her/his own money. Possibly pleasure or joy as well as comfort can have an influence on whether someone, in the eyes of the producer of a product, will be a consumer or user.

However, although the two terms user and consumer can overlap each other, it is important to observe the differences between them, since market rhetoric has to be designed differently, depending on the form of usage that is intended for the product. Also, it is extremely important to distinguish between customers and users. Customers are to be focused on by marketers and sales personnel. To product developers, however, users or consumers, and the use of the products are of prime importance. As pointed out earlier, product developers also need to consider that users can be divided into primary users (i.e., principal users), secondary users (i.e., temporary users), and co-users (i.e., people involved but not using the product themselves).

If it is the goal of a company to remain on the market for a long period and to be economically sustainable, then a must is that the customers are content, not only at the moment of purchase, but throughout the practical life cycle of the product. Good customer service, well-designed manuals (printed and in digital form!) and environmentally adapted/recyclable solutions are some examples of important issues for the company in this respect. These factors affect whether a repeat purchase will take place in the short term and whether the customers/users are ambassadors or influencers and contribute to the spread of favorable judgments on the company and the products.

Research shows that customer loyalty has enormous significance to a company's long-term profitability (e.g., Carr 1999). Common knowledge and research also show that a customer/user may be "loyal" to two or more companies at the same time. For example, in a family there can be two different makes of car to which loyalty is shown when changing cars. Therefore, a better term than a loyal customer is maybe a returning customer.

Customer loyalty is built up successively in what may be termed relational stages in four steps. The four steps, of which only the first counts initially for innovations are:

- Non-customer: the individual has little or no knowledge of the supplier and the supplier's offers, has never bought from the supplier, and has no plans to do so.
- Potential customer: is aware of the supplier and the supplier's offers, and perceives a need for the product.
- Customer: has bought at least once.
- Loyal/returning customer: has bought several times and/or buys most required products from the supplier.

8.4.2 Buyer Categories

Classical marketing literature often states that success in product sales demands a suitable choice of four Ps. They stand for Price, Place, Product, and Persuasion/promotion. Sometimes these four Ps are also complemented with four Os (Object of purchase, Objectives of purchase, Organization for purchasing, and Operations of purchasing organization).

Today we know that neither the four Ps nor the four Os are sufficient for successful marketing and sales. This is, above all, because the crucial human, irrational dimension has been ignored in their models. Marketing via personal influence, where the rational and irrational factors may be constantly influenced and the message thereby amended, is particularly important in selling new products. However, energy and commitment may be devoted to individuals who will not buy the product until they know that it has been previously bought and used by others.

Many individuals often influence the purchase of a product in a complicated manner. For example, Gummesson (2002) has identified thirty relations (30 R) in connection with a purchase. In practice, however, it is sufficient to remember that a deal is often based on relations in many directions, accompanied with influences from other persons, such as active on *social media* and from *viral marketing* (see below). One example of this is that a patient hoist generally has one buyer/customer (the medical authority), many users (paramedics, nurses, patients and relatives) and many pressure groups (doctors, health and safety officers, relatives, and supervisory authorities). All these or other categories affect the company's possibilities to sell its product. Each of all these categories is influenced by their individual networks as well as by social media.

From when sales start to take off, it is important to know the marketing target group as well, as this is where the group is geographically located and this has a major influence on the choice of message. Furthermore, in selecting the target group, it is vital to consider that, with regard to the experience of the buyers, they can be roughly divided into the five different buyer categories earlier described (pioneer buyers, early adopters, early majority, late majority and latecomers/ laggards). The statistical number of possible buyers from the launch of a new product and until the latecomers have bought the product is shown in Fig. 8.2.

Pioneer buyers are often active information seekers about new ideas and are persons who, by nature, often want the latest type of products, which they enjoy displaying. They have, in general, a high degree of mass media exposure, and their interpersonal networks extend over a wide area, reaching outside their local system. They are generally able to cope with higher levels of uncertainty about a product than the other adopter categories.

Pioneer buyers are also a kind of ambassador for their new products, provided that they are not dissatisfied, for example, with the quality and service. They also

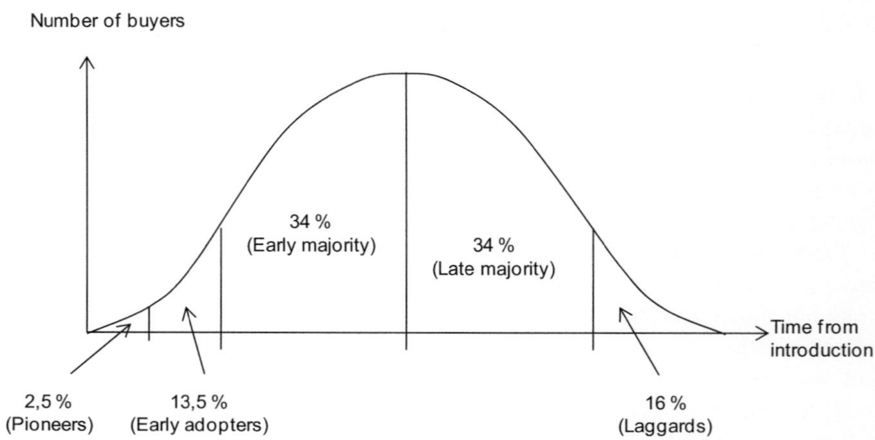

Fig. 8.2 The number of first-time buyers of an article/product (or an opinion) is small at first, and will then increase and peak, after which it starts to decrease

seem to be prepared to pay a high price simply to be first to own their new product and want to show off with it. There are pioneer buyers both for the consumer product category and, for example, among buyers of hardware for a company. Constructers, buyers and secretaries then, can be divided into the different buyer categories. It is important to remember that those who make purchasing decisions regarding a new product are also one or more individuals in the company and are seldom computers although these can be programmed to handle repeat orders, thus act as automated, purchasing robots. Such 'robots' are common in the financial market, for example, in the stock market trade.

Even though each person has a basic mindset that fits one buyer category more than another, we act in different ways with regard to different situations and to different products and opinions. Most people, therefore, can certainly remember being ahead of most or many others regarding some purchases or opinions. In other cases, they have waited before accepting a new product or opinion until others appear to have accepted them. Thus, there are few pioneer buyers who act as pioneers in every situation. However, the probability that a clear pioneer in one respect is a pioneer also in another is greater than that a clear latecomer should suddenly act like a pioneer.

The normal distribution curve in Fig. 8.2 provides a mental picture of what, in principle, happens to a product on the market during its entire commercial lifecycle, i.e., from its launch on the market as a new product until it is removed from the market by the company. In reality, the curve is only very rarely evenly distributed; it is usually biased one way or the other for each individual product. In certain cases the number of pioneers is only a few per cent of all first-time buyers, while in other cases it could be around ten per cent depending on whether the product is sought after or meets a pent-up need.

Due to natural disasters, military and trade conflicts, terrorist attacks, global financial crises, and new laws and regulations, sales can abruptly cease. Even if a new product is in great demand, for example, a vaccine against a disease, and the number thereby can be appraised, it is difficult to estimate time-wise how sales will take place in reality, as marketing and sales are merely two influencing factors. Other important factors are, for example, the purchasing power of the target group, the economic climate, weather conditions, competitors' countermoves, sales channels, or political measures. Extensive business planning before market launch is therefore often wasted time and money.

As a consequence, the important point to bear in mind regarding Fig. 8.2 is that there are different buyer categories with different preferences, and buying behaviors. Without this insight, marketing and sales will be ineffectual. An entrepreneur, who thinks that his company can skip the pioneer buyers and aim the drive at early majority buyers, is not likely to be especially successful. However, the time it takes to reach new buyer groups differs. One example of super short time is the spreading of Pokémon into different categories of users.

Even if one wants to reach the early majority group of first-time buyers quickly, it is still not possible to reach them without first having won the pioneer group. For this reason, it is often an erroneous approach to carry out price and perception control of a statistically selected group before the launch of a new product. Tests like this will show a forced situation where time is frozen at a stage later than the initial stage. When sales then take place, time is moved back to point zero. If one goes by statistical tests, then one's aim will usually be wrong. This is because one then tries to sell the new product to the majority group of buyers, which, time-wise, is not prepared to buy the product. At the same time, the pioneer buyers will not want the product, since it is not made for them. Although one feels that one has done everything right, sales are often poor if faith has been put in this type of statistics and statistical surveys.

The only initial interest one should nurture when marketing new products is in finding out how some pioneers perceive the product. If possible, before developing and manufacturing the product, one should also sell the product concept to a single representative of these pioneers. Thereby, the product will be used and evaluated at an early stage, and this will help the product developers to focus their efforts on producing a product that is attractive to the buyer.

If the company, salesperson or pioneer buyer gives away the product to a user/consumer who mentally belongs to a later user category, this user will often not use the product voluntarily. Worst of all is if the pioneer buyer gives the product to a user who is a latecomer by nature. One can then be practically sure that the product will not be used.

Coming back to the example, if a seller of treatment equipment does not find one or several pioneer users among care personnel, the product will remain unused no matter how good the seller's contacts are with the medical authorities' buyers. In this case, it is often necessary for the seller to interest a "pioneer patient" in the product instead and one who dares to test it. It is also important to ensure that health and safety officers and doctors form a favorable opinion of the product. To establish a new product successfully means putting the focus on finding and linking as many pioneers as possible among buyers, users and other influential people.

8.4.3 Attitude Development

The attitude we have about a product and the provider is critical if we are thinking of buying or re-buying it and, if so, what is an acceptable price. This individual attitude is based on many things. It is also time- and situation-dependent so that a positive attitude can become negative for one reason or another. This change can be rapid, for example, if an accident occurs with the product, or slow if a design is not changed for a long time.

Our attitude about a product is based on the experiences we get from the use of it, how it "affects" us and what we hear about other users' experiences and affects (affect = experience of feeling or emotion). Next in line is information from "independent" sources. Also marketing material will affect our attitudes although, in general, we may be skeptical about such information. This is why some sources say

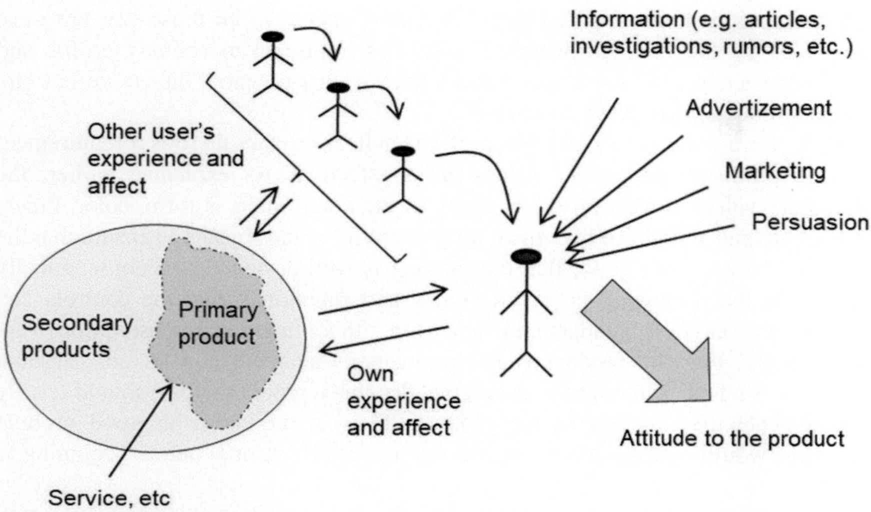

Fig. 8.3 Gradually our own image of a product is developed by input from different sources

that a positive article in a newspaper is up to twenty times more efficient than an advertisement.

Thus, the total attitude we gradually build up in our unconscious minds is dependent on many inputs (see Fig. 8.3). This individual attitude becomes rather clear when we think of the product not being directly influenced by it, for example, when we close our eyes and do not taste, touch, smell or hear it.

One psychological theory, the lateralization of brain function, holds that when we build up an attitude to a product, one half of the brain deals mainly with the affective or emotional feelings, while the other half deals mainly with the cognitive or rational reality. Therefore, emotions (passion) are contrasted with cognition (reason) as a source of motivation and decision making. However, the term cognition is used in different ways by different disciplines. In psychology, it refers, for example, to an information processing view of an individual's psychological functions. Often cognition is interpreted as understanding and trying to make sense of the world. In reality, an individual's emotion, cognition, and behavior are interwoven in a complex manner.

Everybody has probably experienced how our attitude to a product can be substantially influenced by our emotions—our passion—about it. Therefore, being passionate can cause us to buy a product initially without having any logical arguments for such a decision. Afterwards, we often build up logical arguments for the decision. As the emotional values are difficult to influence, the difficult job of every seller is to please the buyers and users as much as possible after the delivery of the product so that they become good product ambassadors for the product. As buyers, we act more or less rationally, depending on what feelings we have regarding the product. Our choices are steered by irrationally and/or rationally-based motives. The biggest decisions appear to be those that are most influenced by our feelings, which is why it is important to actively profile and protect one's company name and brand names so that potential buyers do not buy similar products from other companies.

Thus, for a purchase to take place of a technical product there is a requirement that the product's perceptory values are satisfactory. As explained earlier, the perceptory values consist of one or many of the items, such as form, color, finish, smell, taste and sound. To maximize the perceptory values means to ensure that the product semantics are good, that the product is well-designed and ergonomically made, that the packaging is well-designed and functional, that the controls and buttons are easy to understand and use, that functional descriptions are user-friendly, that the product is environmentally-adapted, and so on. [Product semantics is a field of thought based on the idea that a product's form should readily communicate the function of the product. Thus, a product with good product semantics would typically not require the user to read a manual before beginning to use it.]

However, without good functional values of the product, a sustainable sale will not be achieved. When that first step has been achieved in the PD process, all of the product values mentioned in Chap. 2 need to be considered.

8.5 Marketing Concept Development

8.5.1 Market Research

Before a new product is marketed, one has to have an opinion on who must, want, or will buy the product or service. To get information to build an opinion on in a simple way, the start can be to convince someone to buy the product or service and then to learn from the trials who actually can be customers and what to communicate to them. A more sophisticated way is to first carry out market research and then to develop and test the marketing strategy in reality, which is the way larger companies work. In that case, market research is an important component of the total business strategy.

More specifically, *market research* is meant to be any organized effort to gather information about target markets or simply customers (McQuarrie 2005). It is often interchanged with the term *marketing research*; however, marketing research is concerned specifically about marketing processes, while *market research* is concerned specifically with markets (Mc Donald 2007).

Market research has the aim to provide important information to identify and analyze present market need, market size and competition as well as possible changes to come. Market research techniques can encompass qualitative techniques, such as forming and getting information from focus groups, making in-depth interviews, and exploring cultural phenomena (ethnography). Market research techniques can also encompass quantitative techniques, such as making customer surveys, and making an analysis of secondary data. It also includes social and opinion research, which involves the systematic gathering and interpretation of information about individuals or organizations using statistical and analytical methods and techniques of the applied social sciences to gain insight or support decision making (ICC/ESOMAR 2008).

Based on simple or elaborate market research, a marketing strategy can be made (The term *strategy* means a high level plan to achieve one or more goals under conditions of uncertainty. The term *tactic* is a conceptual action implemented as one or more specific tasks).

8.5.2 Marketing Strategy

Marketing strategy has the fundamental goal of increasing sales and achieving a sustainable competitive advantage. Marketing strategy includes all of the basic short term, and long term activities in the field of marketing that deal with the analysis of the strategic initial situation of a company and the formulation, evaluation and selection of market-oriented strategies. Therefore, it contributes to the goals of the company and its marketing objectives. Marketing theory should focus on the value of the total offering, a balance between production-centric and customer-centric aspects, and should recognize relationships, networks and interaction as core variables (Gummesson 2002-B).

The process of making a marketing strategy usually begins with a scan of the business environment, both internal and external, which includes understanding strategic constraints (Baker 2008). It is generally necessary to try to grasp as many aspects of the external environment as possible, including technological, economic, cultural, political and legal aspects. Goals are chosen. Then, a marketing strategy or marketing plan becomes an explanation of what specific actions will be taken over time to achieve the objectives. For large companies, plans can cover many years, with sub-plans for each year, although as the speed of change in the merchandising environment quickens, time horizons become shorter and shorter. That fact underlines the necessity of using dynamic principles and mind settings for every company and organization so that it can become sustainable.

Ideally, strategies are both dynamic and interactive, partially planned and partially unplanned, to enable reaction to unforeseen developments while trying to keep focused on a specific pathway. Simulation activities, such as using customer lifetime value models, can help marketers conduct "what-if" analyses to forecast what might happen based on possible actions, in order to gauge how specific actions might affect such variables as the revenue-per-customer and the churn rate.

8.5.3 Brand and Branding

The noun *brand* is a recognition device/attribute of a company or an organization. It is a combination of many supporting parts, as a name, a term, a design, a symbol, a music or acoustic piece, or any other feature that identifies one company's or organization's products or service as distinct from those of others. *A brand gives a set of expectations, memories, stories and relationships that, taken together, account for a customer's decision to choose one product or service over another.* The brand is ultimately what determines if someone will become a loyal customer or not. The marketing may convince someone to buy a particular product, but it is the brand that will determine if that person will be a loyal customer or not. Therefore it is important for a brand's identity to be visually consistent everywhere people see it. It has both a look and a feel. The look is its visual identity. The feel is its emotional identity.

A brand is built up by different branding efforts. Thus, the verb *branding* means actions to create a brand, to build up an image that people recognize and connect to in a positive way, independent of whether the brand is connected to a company or an organization such as a governmental authority or a club.

A *brand strategy* is how, what, where, when and to whom to make the branding activities. Where to advertise is part of a brand strategy. The distribution channels are also parts of a brand strategy as well as what to communicate visually and verbally. The use of well-known persons as famous bloggers, actors, sportsmen, musicians, and scientists are other examples of parts of a brand strategy.

Two central parts of a brand are the name and logos used for different products and services. The name should be easy to remember and be useful in many languages, if the products and services are to be sold in more than one local market. A logo is the graphic symbol that can be purely typographic. When only the letters of the name make up to the logo, it is called a logotype or wordmark. Some people also refer to the logo mark as the word portion of a logo that also has a symbol. Sometimes the graphic symbol and typographic word mark are very separate. With other logo designs, there is not a clear separation of the logo symbol from typography.

The logo helps customers discover, share and remember a company's brand. A good logo reflects the mission of the company or organization and is:

- Simple
- Memorable
- Timeless
- Versatile/universal
- Appropriate/hitting.

An example of the process to make a brand is that, when we started a company with innovative products for disabled people in 1990, we first created names for people to connect to and to remember, independent of place in the world. We settled on the name Handiquip, although that "h" is not pronounced in the French

HANDIQUIP

Fig. 8.4 The final logo of the brand Handiquip®

language. Handiquip is a combination of the two words *handicap* and *equipment*. Next step was to make a logo that supported the image of the business and its name, which became possible by extending the Q to be like a supporting hand (see Fig. 8.4). As the company was a Swedish company, the colors of the Swedish flag were used in the same way as IKEA® has done. Now a complete graphic profile could be done and was registered as a trade mark—Handiquip™. The company is still in action (www.handiquip.se). Note that the name had been questionable to establish, as to be handicapped is negatively loaded. This is why the word disabled is used instead of handicapped.

8.5.4 Content Marketing

The content in marketing has the aim to describe the company or the organization in a positive and trustworthy way. The content can be the history of the company or organization, ethics and morale, beliefs, mission, size, and geographical markets where the company is present.

Based on the content, content marketing is a strategic marketing approach focused on creating and distributing valuable, relevant, and consistent content to attract and retain a clearly-defined audience. Basically, content marketing is the art of communicating with customers and prospects without selling. The intention is to change or enhance customer behavior. It is an ongoing process that is best integrated into the overall marketing strategy of the organization.

To take one step back, the noun *content* has different meanings to different people and situations, depending on the context they are in. For some it means information. For others it means experience. In business terms, content is the information a company or an organization provides to their defined audiences and potential customers. It is through the provided content that people discover, consume and act on brand information. All marketing, independent of type, is based on content.

A useful definition of content is (Wikipedia 2016): "In publishing, art, and communication, content is the information and experience(s) directed towards an end-user or audience. Content is "something that is to be expressed through some medium, as speech, writing or any of various arts". Content can be delivered via many different media including the Internet, television, audio CDs, books, magazines, and live events, such as conferences and stage performances."

A simple definition of the term *context* is (http://www.businessdictionary.com/definition/content.html):

- the words that are used with a certain word or phrase and that help to explain its meaning,
- the situation in which something happens: the group of conditions that exist where and when something happens.

The content is based on a *content strategy*, which refers to the planning, development, and management of the content, written or in other media.

In the digital world, a *content management system (CMS)* is a software application used to upload, edit, and manage content displayed on a Website. This software also enables less technical individuals to manage content on a Website easily without having an extensive coding background.

Basically, *content marketing* is the art of communicating with customers and potential customers (prospects) without selling. It is non-interruption marketing. Instead of pitching products or services, content marketing is delivering information. The essence of such a *content strategy* is the belief that delivering consistent, ongoing valuable information to buyers, they ultimately will reward by being loyal customers.

Based on Wikipedia's (2016) popular descriptions, common forms of content marketing are:

- *Social media marketing*: Social media are computer-mediated tools that allow people, companies and other organizations to create, share, or exchange information, career interests, ideas, and pictures/videos in virtual communities and networks.
- *SEO—Search Engines Optimization*: the process of affecting the visibility of a Website or a Web page in a Web search engine's unpaid results—often referred to as "natural," "organic," or "earned" results.
- *PR—Public Relations*: the practice of managing the spread of information between an individual or an organization (such as a business, government agency, or a nonprofit organization) and the public.
- *PPC—Pay per click also called cost per click (CPC)*: an Internet advertising model used to direct traffic to Websites, in which an advertiser pays a publisher (typically a Website owner or a network of Websites) when the ad is clicked.
- *Inbound marketing*: the promotion of a company or other organization through blogs, podcasts, video, eBooks, newsletters, whitepapers, SEO, physical products, social media marketing.

8.5.5 Market Segmentation

A classical marketing view is that successful marketing is based on the ability to identify, analyze and develop a few attractive market segments. To identify possible customers and market segments means to find interesting portions of the total market that can be targeted. Judgment points are future possible market size, geographical location and one's own possibility to meet expected requirements, after which one segment is chosen on which to concentrate all marketing and sales efforts.

However, this analytical way of working is often unsuccessful for innovations. Our experience tells us that one should use a "shotgun" principle instead of sharpshooting at only one market segment. The sales people of a new solution, therefore, must try to get a customer, independent of the ranking of interesting market segments. When having got one or more customers, the marketers can use analytical methods to take better care of that market segment. Meanwhile, or in parallel, the sales people should try to sell to other market segments with the aim of getting a number of segments growing. In time, often after three to seven years, history will tell which segments to concentrate on and which to put on the backburner.

For example, in 2003, the Web specialist company, Tervix ICT AB, managed to sell a number of Intranet systems to organizations that partly got money from the EU. Internally in the company, there was a view that it should concentrate on that market segment. Two years later, the EU stopped further financing of such activities, not because the products or service were bad, but because the EU targets had shifted. Had the company not continued to work on many segments, it would have ended up in a severe situation. Having all your eggs in one basket is, no doubt, a risky strategy especially for new companies selling new products!

A suitable method to use in the creative process of finding possible customers and market segments to approach is to start up dialogues with different people on the issue. For example, relaxed dialogues in small groups over a cup of coffee can be rewarding for creating practical ideas. Conversations (dialogues) with users, and especially lead users, customers and experts are also important possibilities as are meetings in different networks. If this still does not produce results, then brainstorming is a possibility, which, of course, can also be used in parallel with the different dialogues.

It is safer if one can choose a customer/user to co-operate with when the product development is done than if the development happens in isolation from the market. Further, a demanding customer should be chosen if that is possible. However, in an intensive exchange between developers and a demanding primary customer/primary user, conflict situations can arise which may be perceived as difficult to handle. Differences of opinions are, in general, extremely useful to the product development process. Also, customers/users that create demands without making suggestions for solutions are of lesser value than customers/users that create demands but also suggest solutions.

It is worth remembering that customers/users/consumers are not always aware of their "real" needs. In certain cases, for example, for small children and some disabled and/or the elderly, there are little or no communication possibilities. In turn, this means that company representatives must have ample experience of their own to be able to make correct assessments.

8.5.6 Marketing Material Considerations

The most important question for any marketer regarding the marketing material is to answer the question "what´s in it for me?" for a real or considered customer. Then, during the work, to make convincing and interesting marketing messages and material one needs to find out some different facts. A simple way to do this is to make what is the earlier explained as a SWOT analysis, which means finding out one's own strengths, weaknesses, opportunities, and threats. Sometimes, a Pugh matrix can also help to focus on messages to use. Figure 8.5 shows a more thorough map to find out facts of importance for the development of efficient marketing material for which SWOT is only one part.

As explained, the marketing material includes everything from logo, business card, Website (home page), brochures, flyers,—essentially everything visual and outspoken that is connected with organization's name. There is a simple rule-of-thumb for how professionally-produced materials should look. "Create materials to the level that reflects how much you're asking clients to invest in you."

Today, an absolute demand is to have a professional mobile friendly home page and maybe also a Facebook page as people start finding out about the company by going to its Web page and/or smartphone App. This home page must be updated frequently with major revisions at least every year, even for mature businesses.

The marketing material should be unique for every organization but with similarities for product and service families that the organization supplies.

We should remember that the product and/or services an organization markets is not merely the core product, but everything that the customers perceive around the

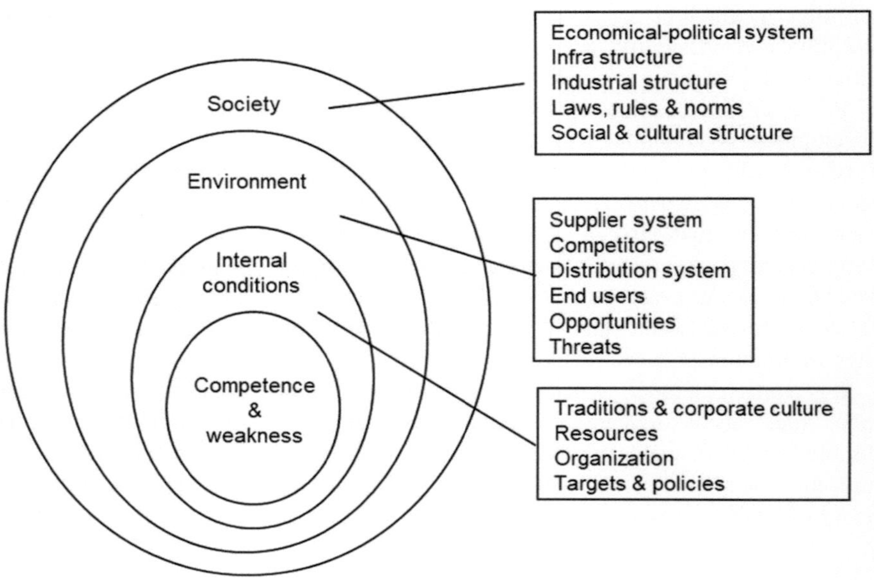

Fig. 8.5 Some internal and external factors to consider for the making of marketing material

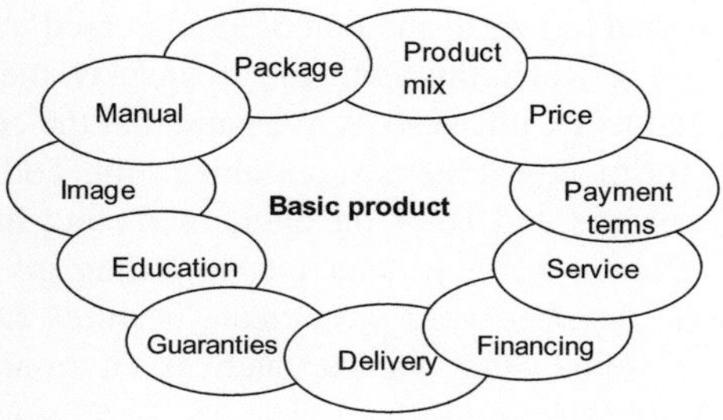

Fig. 8.6 When the customer is placed at the center of marketing activities, the basic product becomes only a part of the total offer

product and which noticeably affects the decision to purchase. Therefore, marketing aims to make an overall offer to the customer, so that the company thereby achieves optimum short-term and long-term profitability—something that can often be seen as a contradiction. Figure 8.6 shows an example of the complexity in the composition of different marketing activities.

It is important in all selling and marketing activities to make a positive and lasting impression on one's contacts. As all employees in the company are marketers to some degree, it is a small cost to give them their own business cards, as leaving behind a business card creates a lasting impression.

A business card must show the person's name, the company name, address, phone numbers, Web address and e-mail address. If the Web address is long also a QR code (see below) is a piece of good advice to print it on the card. The font and general design should be selected for the greatest possible readability. One should, however, avoid advertising on a business card as the information gets quickly out of date. One should also avoid folded business cards as it is usual for recipients to insert them in the plastic sections of a wallet. For the same reason, one should choose a standardized format for the business cards.

As selling new products is largely a question of "selling yourself", it is recommended that at least the sales people each have a photo of themselves on their business cards. Also, it is for psychological reasons more difficult to scrap business cards with photos on them than cards with no photo.

8.6 Pricing Considerations

Classical marketing was basically a question of price and demand, i.e., how is the long term to optimize one's income from the product or products that the company provided. The change of ownership of the product(s) in each case should take place in a manner as advantageous as possible to the seller, while the buyer should perceive the value of the product to be greater than the agreed total price. This gap/difference is sometimes called "metavalue".

When companies started to face a situation of an increased number of competing products, marketing got a wider mission, i.e., to display the product, to create interest, to persuade, and to set prices so as to ensure that the company was able to exchange its products for money at the best possible profit. Today, when we have a surplus of products, services and labor on offer, marketing has to do the job of securing profitability from a wider perspective, including giving away the basic products free of charge. Another word is to create business sustainability, which, however, also includes some other values, such as environmentally acceptable solutions, or ethically right decisions.

A trend is that the basic price is going more and more in the direction of zero and that revenues have to be taken out in other ways, such as in support, service, added functionality, or advertisement. For software, this trend is especially strong and a lot of software is of the open source type, meaning that it may be used free of charge. Advanced mobile telephones are sold at almost nil if the customer signs a contract for one or two years with a telephone company. Therefore, to an increasing amount, money must come from things related to the basic product.

If the company does not sell the product to the end customers itself, it is important when price-setting to ensure that sufficient scope is left for the retailers' mark-up. It is usual for prices to double at each middle step between the manufacturer and the customer. Since one cannot have one price locally and higher prices on other markets, there are contacts between both buyers and retailers on different markets and Web technology is pushing this development forward. Thus, an end price must be applied that is fairly uniform in all markets. Upon launch in the home market, one must use the price that the planned retailers should use. Thereby, much-needed margins for the business operation are created in the initial phase.

Price-setting of new products is difficult and multi-dimensional. Price-setting of ground breaking new products is extremely difficult as there are no or few other products to compare prices with to get an idea of acceptable price levels. Other reasons are that, until a few new products have been sold, there is no generally accepted price level to relate the price to, and there is little or no information available concerning price-sensitivity. As there is no information available concerning competitor reactions before the new product has been launched, there is no knowledge concerning customer reactions to the new products before they have been introduced to the market, and, therefore, there is a lack of knowledge about which marketing and sales arguments give the best effect.

When discussing prices, some basic knowledge is useful. The first is that, in general, we are not prepared to pay much for basic products, while we can pay extreme sums for what are sometimes called premium products. Using Maslow's stairway (see Fig. 8.7) for guidance, we will find premium products are high up on the stairway while the basic products are low on the stairway.

Next, the possibilities to get high margins for products depending on the emotional values can be pictured as shown in Fig. 8.8. For example, the kilo price for a product with lower emotional and image value, such as for wheat or a Škoda car, is very different from a product with higher emotional and image value, such as a lipstick or a Porsche car.

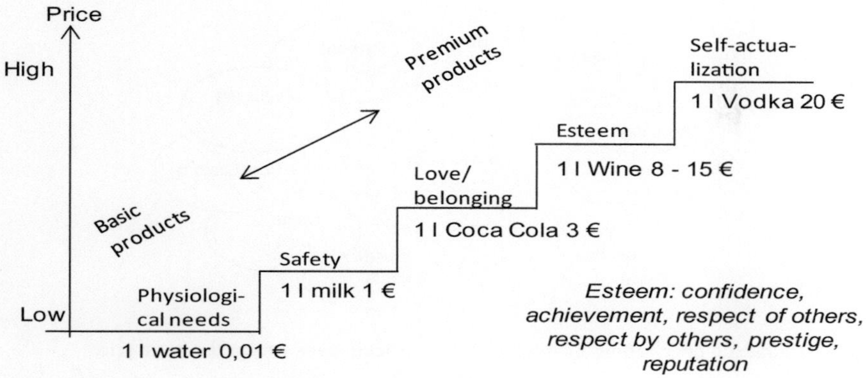

Fig. 8.7 Basic and premium products can be placed in Maslow´s stairway

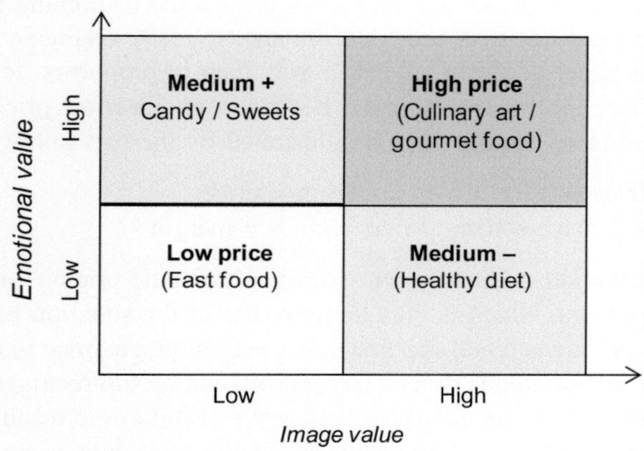

Fig. 8.8 Margins measured in price/production cost varies a great deal, depending on emotional and image value for the food-related products

Then, price-setting will be influenced when a customer is prepared to buy the product. By giving different payment alternatives, the decision time can be moved in different directions. The costs of using the product will also influence the decision as well as service costs, and insurance costs. Altogether, the palette can be called "strategic price-setting". Figure 8.9 shows an example of things that will influence the time from knowledge of the new product to the decision to buy it.

The classical view of setting the basic price to cover all the different activities in a company that is undergoing changes. Instead, when the customers get primary products free of charge or at a very low price, they will often pay for secondary products or for extensions of the primary products and/or services. One of the early success stories of this was Xerox's way of selling their copier machines free of charge, and then letting the customers pay for with every copy that was made. The Tetra Pac Company used the same strategy to get sales going for milk distribution.

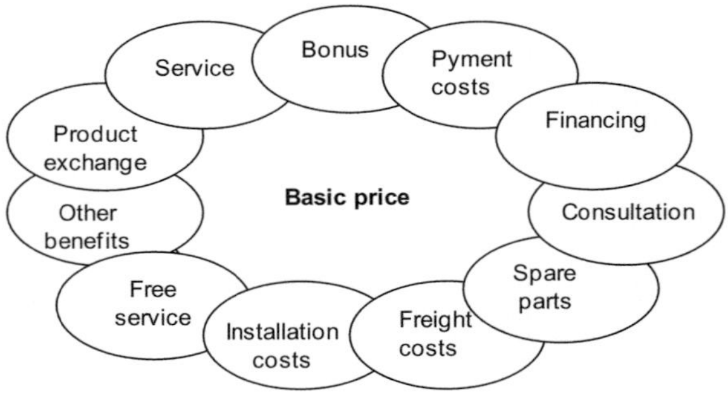

Fig. 8.9 Strategic price-setting includes many influencing parts except the basic price

Classically seen, market-oriented price-setting for hardware products is the price that corresponds to the usefulness of the product and has to have a sufficiently large number of users. Also, the manufacturing cost has little or nothing to do with the usefulness a product has to a user. Unfortunately, many engineers still see it as almost immoral to set prices in any other way than in proportion to the manufacturing cost. The conceptual difference between "engineering price-setting" and "market-oriented price-setting" may be illustrated by the two following models:

Engineering price-setting: Costs + margin = price
Market-oriented price-setting: Price − costs = margin

A variant of market oriented price-setting is "strategic pricing", used when the market is well-known, which is thus the opposite of the situation for innovations. When strategic pricing is done, one first has to set a strategic price to reach the mass of customers. The next step is to set a target profit and by subtracting the profit from the price. Knowing the acceptable cost, different possibilities to reach that cost level are investigated including finding cheap production possibilities, re-designing the product, or changing material in the product. This way of working is, for example, what IKEA is said to be doing.

The biggest product cost is often the manufacturing cost if the product comprises a good deal of hardware. For software-based products, the manufacturing cost is always low or insignificant. If the buyer price that can be set for a hardware-based product is lower than three to four times the manufacturing price, one should, as a rule, not begin to manufacture and sell the product. This mark-up factor, however, is not fixed and varies by product and sector. In pharmaceuticals with a long development period, the factor needs to be very high to cover future R&D. For extremely expensive products, such as aircraft, ships, or energy producing plants, it is often not possible to have a factor exceeding 1, 5, while for products with low prices, one needs a mark-up factor in the region of 5–10. Large companies are often more prepared to pay high prices for services than small companies. Price-setting of software-based products cannot be compared with the manufacturing price at all, as the solutions often are given away. In that case, income is generated, for example, by support activities, consultations, or extra code.

When price-setting a new product, it is important to remember that once the product has been launched on the market, it is extremely difficult to raise an initial price that was too low. Instead of a large price rise, one must make small rises (e.g., 5–10%) over a long period in order to avoid raucous protests from sellers and customers. If, on the other hand, the initial price is set too high, there is always a possibility to reduce the price, and a price reduction is normally greeted positively by the market, which is never the case with a price rise. Experience shows, above all, that sales personnel often side with their customers every time the company introduces major price rises. Thereby, the company encounters an internal conflict with the sales personnel and an external conflict with the customers when prices are increased.

If a number of units have been sold at a high price, which the company nevertheless wants to reduce, for example, owing to the appearance of a competitor with lower prices, it is important not to irritate or disappoint the first buyers by suddenly offering the same product at a lower price. The ambassadors for the product, of which some are the first pleased buyers, will no longer benefit the company in such a case, since they feel cheated. This is serious, since the "jungle telegraph" information on the market is very important for all new products. When reducing the price, the product must be modified somewhat so that the market perceives the new version as a somewhat cheaper product sold at a lower price, while at the same time, the first version should remain on the market mainly in its original design at the same price or a price close to it, or maybe even higher.

For substitute products, which are intended for sale on an existing market, there is relatively little scope for free price-setting. In this case, it is usual to assume that the competitors' price level is the starting-point for one's own price-setting. A small price rise of, for example 10–20%, can often be justified by product and/or user advantages. If this tactic is used, it is more difficult for competitors to take countermeasures. They can scarcely raise their prices because a competitor has entered the market at a higher price!

If, on the other hand, one enters the market at a lower price, there is a great risk that this measure will start a local price war, which can be pursued successfully only by a player with plenty of money or one that has accumulated reserves.

By and large, before the launch of a new product, it is always crucial to consider what the competitors will do when they suddenly discover the new player. A new competitor is seldom welcome on the market by them and in order to spotlight the competitors' potential, one can have some colleagues or friends pretend to be competitors in an imaginary market strategy game. A "game" of this type leads to the unconscious and conscious formation of an action plan to be used when the real market game starts, i.e., when the product is launched on the market.

Personal experiences of the introduction of groundbreaking new products show that the remarkable situation often arises where pioneer buyers are not interested in buying the products if their price is set too low, whereas a high price often increases interest in the products. Pioneers, then, seldom wish to buy a product because it is cheap, but rather because it satisfies a personal want for something unique, or a wish to be noticed. Too low price-setting, therefore, even seems to be able to produce a repellent effect on the pioneer buyers, both in a market perspective and from the point of view of coverage.

When the pioneer buyers have been provided with the new products, then product differentiation is required (i.e., user adaptation) to reach the so-called early majority group. Product differentiation is done, above all, in order that the more sophisticated pioneer version is upgraded for the pioneer group's repeat purchase at the same time as the simpler, cheaper version is launched, which is adapted to the early majority buyer group. If one does not make such a product and price differentiation, the number of buyers can diminish unfavorably.

When the late majority buyer group is eventually to be targeted, the usual procedure is to issue an even cheaper version of the product. At the same time, one must also reduce prices in the rest of the range so as not to lose market share.

Investigations have shown that costs and selling prices for a mass produced product often drop by 70–80% for each doubling of the cumulative sales volume when it is established and accepted by the market. Gordon Moore, the former chairman of the board of Intel, suggested that every eighteen months, the number of transistors be doubled in computer memories without price increases for the customers. His assumptions held for thirty years, which is why one talks about Moore's law. These experiences are important to bear in mind, and emphasize the importance that initial price-setting should be at a sufficiently high level to allow the establishment of business operations with long-term profitability.

When a mature level has been reached for the products, many large companies tend to measure their progress in market share. They then pay careful attention to how their market share changes over time. Through temporary price cuts or other marketing activities, a competitor can, for example, temporarily win market shares on a local market unless the other competitors quickly respond to the challenge. If a competitor is permitted to operate for a long period at a lower price level, thereby gaining a larger share of the market, then customer loyalty towards this company will increase at the same time. If the competitor then raises its prices to the normal level, it will, as a rule, retain a larger market share than before the price reduction. The term "hysteres" (meaning lasting) has been borrowed from physics to denote this effect. The retained higher market share is correspondingly called the "remanence".

8.7 Marketing Methods

The number of useful marketing methods and tools are increasing all the time. Here are some examples of different possibilities to give a picture of the marketing method landscape. On a large scale there are three head methods that can be called *traditional marketing, internet marketing*, and *event marketing*. They are linked to each other as is shown in Fig. 8.10. Event marketing uses different parts from both traditional marketing and Internet marketing, depending on what kind of event is

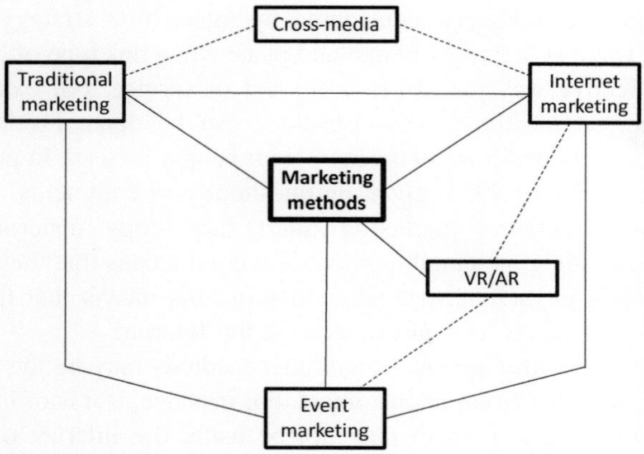

Fig. 8.10 Different marketing methods that are linked to each other

being considered. Cross media is used to approach individuals using both traditional *and* Internet possibilities in a smart way with the aim to visit a special event or to buy a product.

8.7.1 Traditional Marketing

Basically, traditional marketing can be seen as different promotional strategies for which the Internet is not used. It includes a multitude of marketing tactics such as direct sales, TV, radio, mail, print advertising (i.e., magazines, coupon books, or billboards), and printed promotional materials, such as catalogues or brochures (often referred to as collateral). Figure 8.11 shows some of the main methods that are common, depending on what is to be promoted.

Even in the digital age, there are good reasons for employing old-school marketing strategies, including:

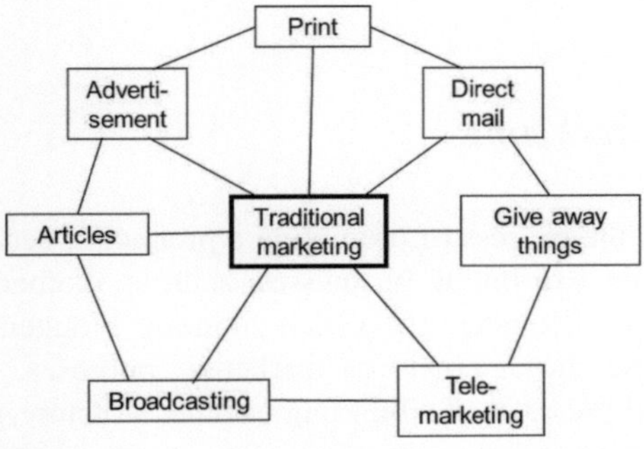

Fig. 8.11 Different traditional marketing methods that are linked to each other

- **Person-to-person selling** is considered by many as a strategy of traditional marketing. There is definitely a time and place when this type of direct selling is the most effective way to market a product or service. For example, through person-to-person meetings, demonstrations can be done. Further, many customers and clients prefer to do business with people they are in personal contact with and direct feed-back is given on questions and comments.
- **Tangibility**. Traditional marketing offers hard copy material. Handing a potential customer some tangible printed material means that they can reflect on the information in their leisure time. In turn, this means that the information often is longer lasting than information on the Internet.
- **Reaching a particular group**. Traditional methods may be the only means of reaching a particular group of customers. For instance, if it is of interest to target retiring CEOs, many of them may not be using the Internet or social media channels.
- **No junk mail**. Many people do not like their mailboxes loaded with "junk" mail. However, it is possible to stop repeated mailings and digital recording has made it easier for television viewers to avoid commercials.

While traditional marketing can be effective, there are some drawbacks or limitations, such as:

- **Expensive and prohibitive costs**: Purchasing advertisements for broadcasting such as TV and radio can be expensive. Printing business cards, flyers, brochures, and mailers also involves printing costs.
- **Difficult to track results**: Traditional marketing can be seen as "throwing things against the wall and hoping they stick". It is difficult to track real quantitative results from the distribution of printed matter.
- **Usually requires outside help**: Printing materials, buying media and creating radio advertisements all require hiring outside help, such as graphic designers or script writers, which adds to costs. However much of that work can, and should also be used for Internet marketing.
- **Forced, one-way message marketing**: Traditional marketing is usually forced upon the users, consumers, and customers, and that can be annoying for them because, for example, they are environmentally engaged or as they feel it disturbing to get a lot of waste to handle.

8.7.2 Internet Marketing

Internet marketing is the process of promoting a product or service online. Internet marketing is not limited to online businesses, such as mothers who blog to recommend products to each other (so called mommy bloggers). Even brick and mortar businesses use the Internet for marketing purposes. (Brick and mortar describes the physical presence of a building or other structure.) Figure 8.12 shows some main methods that are common, depending on what is to be promoted.

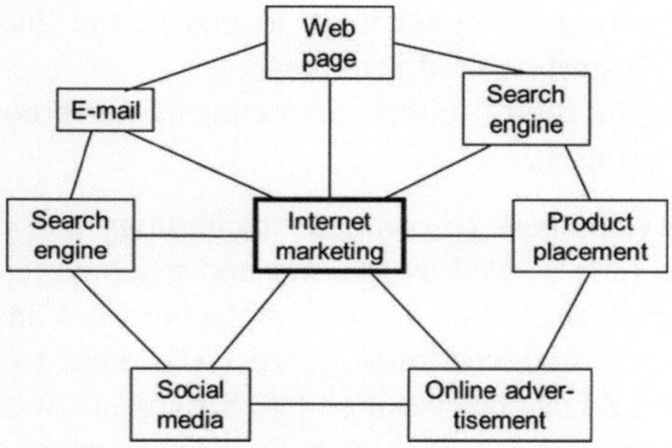

Fig. 8.12 Different internet marketing methods that are linked to each other

Internet marketing uses strategies like website creation, search engine optimization (SEO), banner ads, social media, pay-per-click advertising, and e-mail marketing. Compared to traditional marketing, Internet marketing has made reaching a wider market easier and more affordable. Other benefits include:

- **Results are measurable**. It is possible to find out which methods are leading to sales and which are not.
- **Ability to "drill down" into demographics to accurately reach wanted** target markets. This allows targeting sub-groups for effective marketing messages to increase the response rate.
- **Easiness to make market investigation**. Many online marketing strategies can be easily implemented by a one-person home business. **Decisions can be made using both real data and qualitative results**.
- **Social media benefits**. Direct communication with or individuals and groups of people can be done at no distribution cost. The feedback is quick, for example, on what messages work well and what the market likes.
- **Direct relationship**. The ability to build direct relationships with customers and users via social media and communities is good.
- **Actuality**. The marketing message can be changed immediately when that is needed or wanted. No hindrance of the type "we need to use the printed matter we have before we print new".

While the Internet has made marketing easier, it is not without its challenges, such as:

- **Reliance**. A reliance on customers being highly interactive on the Internet. There are people who do not use the Internet or use it on a basic level, for example, only to send and receive messages.
- **Time demanding**. It can be highly time demanding. Continuous content must be created, edited, approved and published; comments must be responded to and sites and pages must be maintained. Social media can take an inordinate amount of time and become a habit that prevents other important duties to be done.

- **Vulnerability**. There are different ways to misuse the Internet by people and organizations and disturbs sound activities
- **Mismanagement**. Successful online marketing requires consistency across all marketing platforms used.

With the rapid development of computer technology and computer programs, Artificial Intelligence (AI) as well as robotics and machine learning have become important parts of marketing. They are also replacing sales and marketing professionals in online stores. Artificial Intelligence (AI) refers to human intelligence exhibited by machines. AI can be classified into Strong and Weak AI. Strong AI (or artificial general intelligence) is a machine with consciousness, sentience, and mind, and this machine has intelligence in more than one specific area. Weak AI (or artificial narrow intelligence) focuses on specific narrow tasks (e.g., a self-driving car). Machine learning is an approach to achieve artificial intelligence, and deep learning is a branch of machine learning and a technique for realizing machine learning. Deep learning focuses on algorithms inspired by the structure and function of the human brain. Robotics deals with the design, development, operation, and application of robots (Siau and Yang 2017).

Websites can update automatically based on the usage, and Website pages can be reformatted automatically based on eye-tracking data. With AI and machine learning, segment-of-one marketing is possible. The technology can also be used to reveal *dark Web sites*, which are the meeting and market places for emerging cyber threats. These are rich sources of intelligence, often relevant to a broad spectrum of potential targets that are not accessible through conventional monitoring (www.recordedfuture.com).

To end this short summary, some words need to be said about *product placement*. It is basically an advertising technique to subtly promote a product or a service through appearances on YouTube, in films, television or other media (see example in Fig. 8.13). Product placements are often initiated through an agreement between an organization and the media company in which the media company

Fig. 8.13 An example of product placement for Volvo in a children's book (Norlin and Burman 2008)

receives economic benefit. A company will often pay a fee to have their products used, displayed, or significantly featured in a movie or show.

8.7.3 Cross-Media Marketing

Compared to traditional marketing, cross-media marketing uses the possibility of individual marketing. For example, the name of the receiver, independent of whether it is printed on paper or as an e-mail message, will be "printed" where it is strategic in the material.

The material can be communicated by any mass media, such as e-mails, letters, Web pages, or other recruiting sources. This method can be extremely successful for publishers because it can increase the ad's profit from a single advertiser. Furthermore, this tactic generates a good liaison between the advertiser and the publisher.

An example of cross-media marketing can be the recruitment of students to a college. The college can start the marketing process by advertising in newspapers that students read and also shows a QR-code to the home page or App. At the home page or App, the student can write her/his name and addresses together with topics of interest. The college can then, from a modular system, set together relevant information for the student. The next step is to send the material by e-mail and as printed matter to the student. Both the e-mail with a PDF-file and the printed matter will be personalized with the name printed on the material. If an application has not been filed within some time, an automatic reminder card and an e-mail message will be sent. Thus, using both traditional media and the Internet creates a smart way of promoting the student programs and the follow up of the recruitment of single students in a cost-efficient way.

The term "Cross-media" is also used in cross-media entertainment, cross-media communication, and trans-media.

8.7.4 Event Marketing

Event marketing is a promotional strategy that involves face-to-face (F2F) contact between companies and their customers at special events like concerts, fairs, and sporting events. Cross media can be used to build up interest and to invite people to the event. At the event, in marketing entertainment like shows, contests, or parties it is normally used to engage users and customers while they are in a "willing and participatory" position. Figure 8.14 shows some examples of event marketing possibilities.

A successful event marketing campaign provides value to attendees beyond information about a product or service. A discount, free sample, charity alignment,

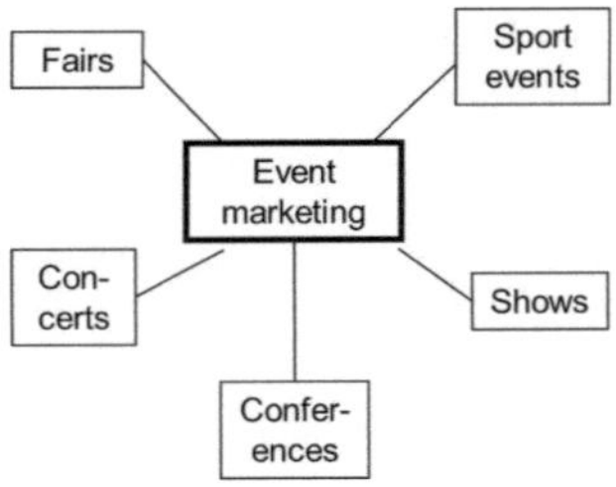

Fig. 8.14 Examples of different event marketing methods

or fun event will make customers feel like they are receiving a benefit and not just attending a live-action commercial.

In contrast to traditional advertising, which blasts millions of consumers with the same general television, radio or billboard message, event marketing targets—with the use of cross-media—target specific individuals or groups at gathering spots, in hopes of making quality individual impressions.

The key to pulling off an effective event marketing campaign is to identify the target audience and to create an experience that remains in the participants' memories. By finding an opportunity to interact with a desired demographic of people—both current customers and prospective buyers—a brand can build favorable impressions and long-lasting relationships. The best, most creative events create interactions that not only reflect positively on the brand at the time, but generate a buzz long after the event is over.

8.7.5 Relationship Marketing and CRM

Relationship marketing seeks to increase sales by building trust and engaging and loyal long-term customers. Relationship marketing is a facet of ***Customer Relationship Management (CRM)***.

The goal of relationship marketing (or customer relationship marketing) is to create strong, even emotional customer connections to a brand that can lead to ongoing business, free word-of-mouth promotion and information from customers that can generate leads. Relationship marketing involves creating easy two-way communication between customers and the business, tracking customer activities and providing tailored information to customers based on those activities.

Customer Relationship Management (CRM) is a term that refers to practices, strategies and technologies that companies use to manage and analyze customer interactions and data throughout the customer lifecycle, with the goal of improving business relationships with customers, assisting in customer retention and driving sales growth. CRM systems are designed to compile information on customers across different channels or points of contact between the customer and the company, which can include the company's Website, telephone, live chat, direct mail, marketing materials and social media. CRM systems can also give customer-facing staff detailed information on customers' personal information, purchase history, buying preferences and concerns.

Thus, while relationship marketing is a sales and marketing concept, CRM refers to the tools used to carry out the concept. Relationship marketing is implemented as a strategy and includes activities such as identifying long-term sales and retention goals, public relations, marketing and advertising campaigns. CRM includes the operational tasks that support the relationship marketing strategy. Activities may include gathering data about the customers and then organizing and analyzing it to create target customer profiles. CRM data is also effective in finding opportunities to create special offers to reward long-time customers for their loyalty, further building the relationship.

8.7.6 Viral Marketing

The cheapest and probably the best way of marketing is if the message about a product spreads like the "jungle telegraph" meaning that a message spreads like a virus by enthusiastic users. This is called *Viral marketing*.

Viral marketing, viral advertising, or marketing buzz are buzzwords referring to marketing techniques that use pre-existing social networking services and other technologies to try to produce increases in brand awareness or to achieve other marketing objectives (such as product sales) through self-replicating viral processes, analogous to the spread of viruses or computer viruses. It can be delivered by word of mouth or enhanced by the network effects of the Internet and mobile networks.

Viral advertising is personal and, while coming from an identified sponsor, it does not mean businesses pay for its distribution. Most of the well-known viral ads circulating online are ads paid for by a sponsor company, launched either on their own platform (company Webpage or social media profile) or on social media Websites, such as YouTube.

Visitors receive the page link from a social media network or copy the entire advertisement from a Website and pass it along through e-mail or post it on a blog, webpage or social media profile. Viral marketing may take the form of video clips, interactive Flash games, advergames, e-books, brandable software, images, text messages, e-mail messages, or Web pages.

The most commonly utilized transmission vehicles for viral messages include: pass-along based, incentive based, trendy based, and undercover based. However, the creative nature of viral marketing enables an "endless amount of potential forms and vehicles the messages can utilize for transmission", including mobile devices.

8.7.7 QR-Codes in Marketing

The QR (Quick Response) code was introduced 1994 by the Japanese company Denso, but not until late 2007 was the first foundation stone for mass use laid when Apple introduced its first smartphone—the iPhone. The next needed foundation stone was the introduction of barcode-scanning apps. Today, QR codes show up everywhere, on almost everything and in different sizes from small squared QR codes to enormous codes that can be scanned from air planes. Figure 8.15 shows such an example.

From a marketing perspective, QR codes offer an obvious value and they are easy to create. The most common uses are Internet links, links to apps, and text. The thought process is that it saves users the trouble of entering a complicated URL.

Some of the methods of using QR codes today make great sense, like on a business card, so people can scan your info instead of having to type it all in. However, the placement of some QR codes makes no sense at all, such as on a street billboard where traffic moves fast.

8.7.8 Virtual Reality and Augmented Reality in Marketing

Virtual Reality (VR) comes from the terms 'virtual' and 'reality'. The definition of 'virtual' is near and reality is what we experience. So the term 'virtual reality' basically means 'near-reality' and entails presenting our senses with a computer generated virtual environment that we can explore in some fashion. Thus, virtual reality is the creation of a virtual environment presented to our senses in such a way that we experience it as if we were really there. Simply, VR gives marketers the opportunity to give their audience the closest experience they can get from a product, service or place without actually physically being there.

From a brand perspective, VR offers the opportunity to make more meaningful connections with the audience. It is used by many of the world's leading brands, such as Nike, IKEA, Volvo, and Redbull to create impactful campaigns.

Fig. 8.15 A corn field cut to form a 29.000 m^2 QR code that can be scanned at a distance of about 3 km. *Photo* Kraay Family Farm

VR allows students to virtually attend centers of learning, museums and galleries around the world. In journalism, VR's ability to see the world through another's eyes allows for new empathetic possibilities.

Whereas Virtual Reality immerses senses completely in a world that only exists in the digital realm, Augmented Reality (AR) takes the real world of the present and projects digital imagery and sound into it. AR builds further on VR by overlaying physical objects with text, images or video.

Whether it is to sell a product or simply have a little fun, there are a whole host of examples of augmented reality being used in retail. Some examples of how AR is used are:

- In the summer of 2013, IKEA launched their augmented reality catalogue to enable shoppers to visualize how certain pieces of furniture could look inside their home. Not only that, but the app measures the size of the products against the surrounding room and fixtures to offer a true-to-life size where possible.
- Many women struggle to find the right shades of makeup or want to know how things will look before they commit to a purchase. An augmented reality makeup mirror first takes an image of a shopper's face followed by showing what the latest cosmetics products will look like on the face.
- Trying on clothes can be pretty off-putting for many shoppers, especially those in a hurry. With AR dressing rooms, shoppers can virtually try on their purchases quickly and easily.
- Choosing the right ring for a loved one can be a difficult task. With AR, shoppers can try a collection through a webcam, and can see how pieces would look in certain lights and against certain skin tones.
- Research by IBM showed that 58% of consumers want to get product information in-store before a purchase, and that 19% of customers will browse mobile devices whilst shopping. To address this consumer need, IBM launched their AR app. Acting like a personal shopper, it uses augmented reality technology to provide shoppers with personalized information whilst browsing the shelves.
- The Converse Sampler iPhone App uses augmented reality to allow shoppers to virtually try on any trainer from their range, simply by pointing their phone at their leg. Customers simply select a shoe from the app's catalogue and see it appear on their foot. Customers can even buy directly thorough the app, meaning they do not need to leave their homes.
- The Burberry Beauty Box store in Covent Garden, London uses AR in a number of imaginative ways. The most prominent is their nail bar. Here, customers can select their skin tone and then place different polishes on the bar. The display then shows how the polishes look in real life.

8.8 Market Channels

8.8.1 General Considerations

There are many ways for a producer to reach the market with a product. The producer can sell the product directly to the customer or via a Web shop and one or many levels (middlemen) between the producer and the customer (see Fig. 8.16).

In the export market, sales can be wholly or partly done via the customer's sales company, or via agents. The Figure includes agents, which are a category that is becoming increasingly frequent as the Internet is developed for direct deliveries from the suppliers. Agents otherwise give information to customers about different products and link the customer and the seller on a commission basis or via subscription fees.

The more middlemen there are between customer and producer, the lower the profit margin for the producer, for the simple reason that there are more who will share the total profit margin. At the same time, volumes in general will be greater with more middlemen when each middleman succeeds in selling well. The income reckoned in cash, therefore, will be greater than at lower sales volumes with a larger profit margin. When selling a new product, before an efficient production and distribution set-up is in place, it is natural—and appropriate—for direct sales to take place on the closest geographical domestic market.

Depending on the state of maturity of the product, the incentive of the customer and seller to bring about a sale will vary. For old well-tried products, such as screws, bearings and engines, for example, one usually says that the customer takes the initiative to buy. For a completely new product, the seller takes the initiative since the product is not widely known. This can be illustrated as shown in Fig. 8.17.

The way the salesman is paid for her/his work influences the speed of the sales and the number of sold products. If the salesman has a fixed salary every month,

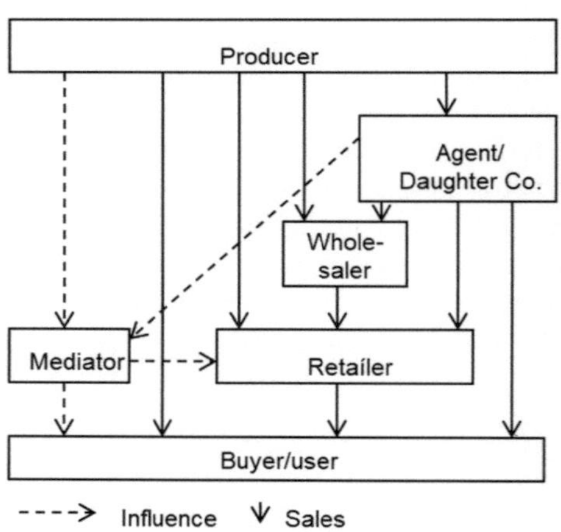

Fig. 8.16 Examples of different options for manufacturers to reach the market

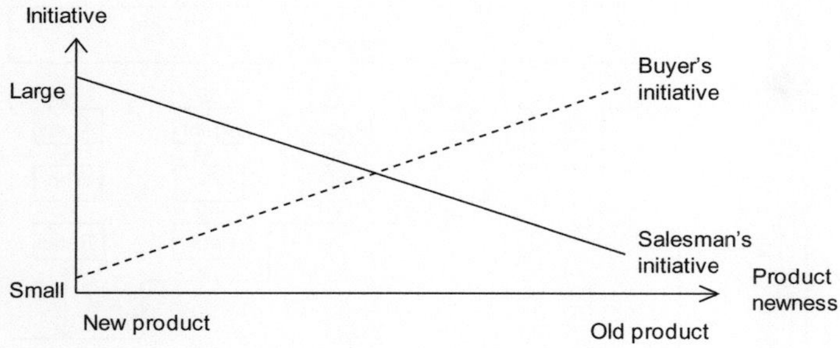

Fig. 8.17 For a mature product, it is mainly the customer who looks for the product, while the sellers look for customers for a new product

maybe she/he does not work as hard as if she/he works totally on commission from her/his sales. A payment principle where the salespeople get a fixed salary and, for example, 10–50% from their own sales produces a situation that is fair for both the company and the salespeople. The higher commission Figures (close to 50% of the sales price) should, in such a case, be used for orders to new customers while the lower numbers (close to 10% of the sales price) should be used for repeat orders. When the time between the first contact and the first order is long—for example for expensive products—a higher fixed salary and a lower commission rate should be used.

Note that the lower the fixed salary the more difficult it is to make sure that the company's ethical code is adhered to by the salesman!

According to an old saying, "all business is local business". The time it takes to introduce a new product on a local market depends not only on the "quality" of the salesman and how she/he is paid but also heavily on how many middle stages there are between the producer and the customer/user. Naturally, the quickest route to the market is when the producer also sells the product. The longest time taken to reach a customer in another country is if one goes through a long series of middlemen. This can be illustrated as is shown in Fig. 8.18.

Some producers strive to reduce the number of middlemen between themselves and the market, regardless of whether they produce mature products or new products. The extreme case of this striving is to completely rely on Internet sales (Web shops) or on TV-shops, and combinations of them. One important reason for making this effort to reduce the number of middlemen between oneself and the end user is that every middleman costs money, with reduced profitability as a consequence, unless the end price to the customer can be raised.

Other reasons are that delivery can be sped up if the number of middlemen is reduced at the same time as the important direct contact with the market becomes faster, as long as geographical distance is not too great. Proximity to the market, then, means that one's products may be more quickly supplied to the customers than without proximity. At the same time, the company can also modify products and marketing methods quickly and in line with customers' changed preferences.

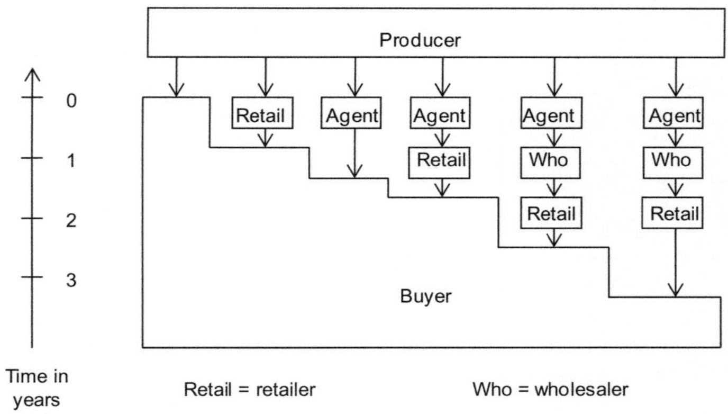

Fig. 8.18 The time to introduction onto the market is based on the number of middlemen between the producer and customer/consumer. The example times are for an industrial product

This also enables better monitoring of competitors' actions and perhaps predictions of when rival products will be introduced onto the market.

The traditional wholesaler feels most affected when producers reduce the number of middlemen between themselves and the end customer/user. One way for wholesalers to avoid problems is to work nearer the market, taking the retailer's role. Alternatively, they can take on a distributor's role, for example through a franchise set-up with local independent partners.

However, for short term reasons, some companies are outsourcing parts of the chain from product development to the delivery and service of sold products. This philosophy can be questioned as when the sales price of products goes down, for example because of the growing competition. The sharing of the decreasing margins with many means profit problems, unless the number of sold products increases enough to cover the sharing of the profit among the different parts.

For a company about to introduce a new product on the market, there is, in general, no better way to start sales than to use the far left possibility in Fig. 7.17. one's own sales geographically close to the company is also a valuable recommendation. Thereby, one reaches the customers/users quickly and receives immediate reactions of immense value to continued operations. However, it is necessary, at a later date, to climb down the steps in Fig. 7.17 so that sales are broadened to new markets outside the local market, which is necessary for income growth to take place. This can be done by the company setting up local offices and/or through joint action with retailers. To reach an export market requires setting up international Web pages and then to start subsidiaries and/or appoint agents. The principle of this growth situation is shown in Fig. 8.19.

When the company has acquired some experience through selling the products on the close geographical market, there is an option for expansion to take place through distributors. Distributors are individuals who act as representatives of the company but who, through their own companies, buy, market and sell the products.

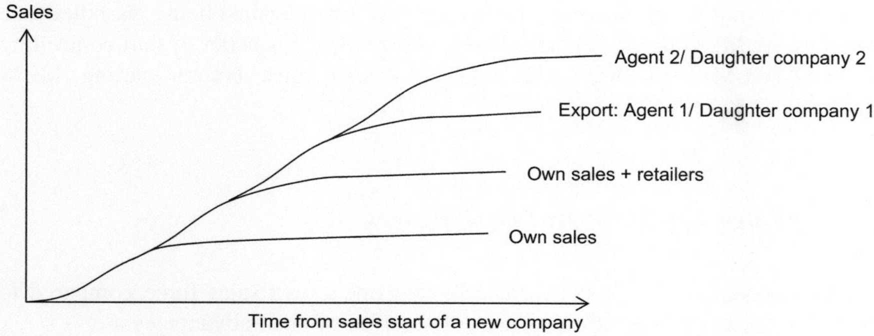

Fig. 8.19 For increased income there is an eventual requirement for new geographical markets to be reached, either direct or via other channels

They, in some way, pay for the rights to do so and have promised to represent the company in ways that have been stipulated. The most well-known distributorship is franchising, which is a business method that involves the licensing of trademarks and methods of doing business. This is, for example, the main form MacDonalds® has chosen for the expansion of their global fast food business. Even IKEA® uses franchising on markets where they have no plans for setting up own warehouses.

Another distributor form that is on the increase is called "Multi Level Marketing—MLM". The idea behind MLM is that individuals act as sales people/distributors without being employed by the company. The products are not sold through other market channels but through distributors, thus giving them exclusive access to the products. To carry out the business, they often form micro enterprises organized in a binary network, meaning that someone above gets commissions on the sales that the lower levels have made. If each new level distributor has to buy a minimum quantity of the products to sell further at fixed prices, then, in most countries, this is regarded as illegal pyramid selling.

For the producers, MLM means that there is no need to set up their own sales force for market expansion. The producer also does not need to pay advertising costs. Therefore, the economic risks are low with MLM as the company will only have a few fixed costs. As the normal competition situation on an open market is not present, the products can also be priced higher or much higher than other equal products sold on the market.

A difficult question with MLM is the ethical/moral effect the system has as the distributors. To be successful, they need to sell the products to people who have confidence in them, or feel sorry for her/him for not selling much—as in the family and with friends. Therefore, they become objects. When they have been convinced or forced to buy they will often have a bad feeling afterwards as they could not say no to the distributor for friendship reasons. If family members and friends are also recruited to become distributors themselves and are not successful in their work, a good friendship can abruptly be broken, eventually leading to social problems, which is not a desired situation.

A sound multilevel marketing company must guard against being classified as a pyramid. As MLM has its longest history in the USA, it is better to start controlling eventual MLM marketing ideas with American rules before starting MLM activities.

8.8.2 Sales via Co-operation Partners

There are a number of advantages to having one's own sales force compared to making sales via co-operation partners. Examples of such advantages are:

- The salespersons can devote themselves fully to selling the company's own products without portioning out their time to other companies. If a situation arises where a choice has to be made, the salespersons divide their time in proportion to the margin that they receive in relation to how easy the sales work is.
- The sales force can be more easily trained and developed to an acceptable skills level, and continuous internal training is made possible.
- The company's own sales efforts can be aimed at tasks that do not give immediate results, such as market studies, the launch of new products, or customer services.
- Key customers expect service direct from the company. The customer, thereby, receives direct information.
- With its own sales force, management receives good reports and feedback about the market and competitors, which permits quick decisions on re-orientation.

However, to build up a sales force requires capital and time, which are not always available. This leads to attempts to run sales through retailers, wholesalers or agents, who, thereby, become close co-operation partners. If trade is not to be done solely via the Internet, the choice of retailers, wholesalers and agents is of major importance to how fast the business will develop.

When choosing co-operation partners, the persons in the organization constitute the most important influence on a positive or negative decision. The size of the company, on the other hand, is of less interest where successful sales are concerned. In addition, one should often avoid the market leader or the second biggest company in a sector. The third company in size, however, usually is "hungry" for growth and therefore provides optimal work input and commitment. However, if one works in a large organization, it can be difficult to admit that it is not the organization and its good financial situation that matters, but the people in the organization, their drive and their incentive.

In choosing retailers, wholesalers and agents in any market, a careful summary of what is required of prospective co-operation partners should be made before beginning the search and the time-consuming discussions that this entails. Even if the desired characteristics vary between companies, the following guidelines are worthy of consideration:

- The co-operation partner should sell her/his present products to the same customer category as for the new product.
- The new product should fit in with the rest of the co-operation partner's range, so that the employees have experience, knowledge and contacts that are directly applicable to the new product.
- The co-operation partner's existing products should not compete with the new product. A limited overlap may be accepted, however, if there are otherwise marked differences.
- The co-operation partner's sales methods and approach to selling should be in agreement with one's own opinions. If one's own business sells, for example, on technological know-how rather than price, then the co-operation partner should do this for the rest of the range.
- The co-operation partner should be financially strong enough to finance her/his outstanding commission and day-to-day costs.
- The co-operation partner should have a good reputation on the market and should be able to refer to previous good results in establishing new products. The co-operation partner must be a serious-minded representative, so that she/he will give the customer the security that is necessary when selling the new products.

The choice of agent, wholesaler or a retailer should never be made without a preliminary detailed financial investigation of the company and a visit to its premises. Personal interaction between the producer and the retailer must also be good to achieve the best results. At the slightest doubt about any of these points, one should try to find another potential retailer, even though this means a loss of tempo in the short term.

When choosing co-operation partners in other countries, the issue is even greater than as described above for reasons of culture, language and other factors. Consultants on the spot are often of invaluable in this work. Since every country has its own regulations and laws, one should always, as a rule, hire legal staff in the country in question when drafting contracts.

To choose co-operation partners takes time and means a risk if one does not already know the persons well. If personal interaction is not right or if one has a feeling that problems of one sort or another can arise, without being able to put a finger on anything specific, one should definitely choose another employee or co-operation partner. Unfortunately, one can never have everything one wants, and must often compromise on wishes and wants. For example, one may wish for a co-operation partner in a strong financial position, but this usually means someone less "hungry".

Finally, something should be said about the problems of operating in a large organization. To refuse to accept parts of the mother company's sales organization for selling a new product is easier said than done, but even so, one should put forward this view if one feels it necessary. The only thing that counts in the end in such organizations is the financial result. Consent to rules that do not work for small businesses hinders good results, to the disadvantage of both one's own business and that of the mother company. In addition, competition is always valuable to the sales organization of the mother company if one chooses another sales channel.

8.9 Feed-Back Investigations

Entrepreneurs and managers need to get improvement feedback from the field with the aim of enhancing the cash flow and/or to acknowledge the staff for good work.

The improvement feedback of the marketing and sales organization can be arranged in different ways. The most natural, cheapest, and fastest way is to act oneself as a potential customer checking out home pages and marketing material, and by listening to how the team members answer telephone calls and/or how they act when in contact with other people. A second way is to let external people do the same work. A third way is to ask potential or real customers for their impressions and experiences as a result of their contacts with the company and its sales organization, which is done by interviewing them or sending a survey to them by post, e-mail or by using a Web-based survey. A fourth way is to let someone act as a potential customer without telling the organization about who will act as "mystery shopper" and where and when this person or these people will do their secret investigations.

For a company to keep the users happy so they will not to spread bad rumors about the products and/or the company, the attitude and work done by the service organization is of utmost importance. To check this, management can use the same improvement feedback arrangement as for checking up on the sales organization. However, instead of using mystery shoppers, they can let some users of the products contact the service people to repair a product with prearranged errors. Doing one's own rigged tests is far better than being faced, for example, with a TV team having made similar rigged tests maybe with the aim of blackmailing a company. Below are some details about different feedback methods.

8.9.1 Observations, Dialogues, and Interviews

Observational studies are used to get feedback on how associates act and how potential and real customers and users view a company and/or a product. Observations allow the investigators to study the associates and the users in real contexts as well as in laboratory experiments. The observers can either be silent, only observing, noting and recording, or they can be communicative during their observations.

Normally, observations and so called ethnographic studies are used to study life in society, but they are also used to study working life in companies, which does not imply living with the staff but being present when the staff members are working. In such cases, the investigator observes what is going on, records meetings, makes tapes, records videos, takes photos, makes notes, interviews people, collects documents and initiates dialogues to ascertain that she/he has understood things correctly.

The difference between dialogues and interviews is that in a dialogue one is free to speak with another person, while in an interview one mainly follows a list of questions prepared beforehand. In both cases, the person engaged to get information can follow up on statements and answers that seem to need clarification. The information written down is biased by the view of the investigator, meaning that it is based on her/his view and experience of life. When simple questions are used and many people have taken part, the answers can be statistically treated to detect patterns.

Some disadvantages with these types of investigations are the risks of not seeing things with fresh eyes and the risk of assuming too much when interviewing, and as a result, not probing deeply enough. The investigator may think she/he knows the answer and does not expose his thinking to alternative reframing. These factors pose considerable challenges to the investigator and require rigorous introspection and reflection. In addition, there is the tricky job of handling organizational politics and loyalty issues towards key persons and the company as the whole.

Something that has also been found and that is worthy of note is that, especially when performing telephone interviews, interviewer intonation may affect the outcome of yes/no or agree/disagree surveys and so produce a larger number of positive responses than expected. Further, interviews can be problematic since: "… detailed analysis reveals the interview not simply as an opportunity for knowledge to be transmitted … but rather as an interactional accomplishment in which knowledge is constructed by interviewer and interviewee during the course of the interview. Interviewers are no longer simply conduits for answers but rather are deeply implicated in the production of answers" Schneider (2000).

However, some advantages with this type of investigations are:

- There is a minimum risk of losing valuable information/data due to forgetfulness or incorrect reconstruction. When reconstructing past events, there are risks of misunderstandings. The investigator has no opportunity to consider the circumstances outside or inside the studied process that may have influenced the result.
- Firsthand information eliminates the influence of other people's understanding of the situation and their ways of expressing it.
- Opportunities exist to rapidly correct interview manuals or to clarify misunderstandings between the questioners and the respondents.
- Unspoken information that is extremely difficult to take into consideration when using classical research methods is captured naturally.

8.9.2 Surveys/Questionnaires

If statistical results are required for some reason, quantitative investigations are used based on questionnaires/surveys. Such methods rely on the identification and election of a few variables out of a myriad of possibilities. The results are statistically processed and presented as numbers in the form of tables and diagrams. The investigator is, in this case, isolated from direct, personal observations of activities.

The answers from those that care to answer the questionnaires are analyzed with statistical methods. It is commonly felt that this safeguards against subjectivity, which is questionable for different reasons, as sayings such as "a fool with a tool is still a fool" or "garbage in, garbage out" try to tell us.

However, a questionnaire can be a useful data collection tool when the following conditions are met:

- The target audience can be clearly defined and identified.
- The majority of respondents know what is asked of them.
- The focus of the analysis is numerical, i.e., the questionnaire yields quantitative data.

It is claimed that with careful planning, questionnaires can achieve good response rates and provide anonymity; the latter may encourage more honest and frank answers than interviews.

One apparent weakness of the quantitative survey method compared to more qualitative methods is that theory in use, and other information, might go totally undetected because the investigator is isolated from what is really happening. One simply has very little basis for a sound decision regarding how to improve the situation.

Although well designed:

- Surveys/questionnaires are blunt instruments, as only that which is asked will be answered.
- The investigator must second-guess the respondent and very little qualitatively new information is gained.

Other disadvantages are:

- The investigator generally has no idea if the questionnaire was filled in by the respondent it was meant for.
- If there are confusions caused by the questionnaire, the investigator cannot clarify them.
- There is little flexibility for respondents to present their own perspective on issues unless there are several open questions.

Questionnaires can give rise to very poor response rates, yielding non-respondent bias. In general, the lower the response rate the higher the probability of non-response error. Poor response rates can lead to bias because some groups are less likely than other groups to fill in a questionnaire. A non-response rate of higher than 10% may bias the results (Marshall 2005). If direct mail surveys are used, every effort should be made to obtain returns of at least 80–90% or more.

There are several tactics described in the literature as to how to increase response rates:

- simplification of the questions
- shorter questionnaire
- small incentives, such as movie tickets
- the use of stamps both on the outgoing and incoming mail
- paying out money beforehand, which is much more effective than paying afterwards.

8.9.3 *Mystery Shopping*

Mystery shoppers engage front line employees in everyday customer service situations to provide management with a true, unbiased measurement of customer service and experience. Through their eyes, the managers can understand what their teams do well and can build on identified strengths. The mystery shoppers can also uncover hidden deficits where training sessions would be more beneficial. They visit stores, restaurants, banks, salons, movie theatres, hotels and other businesses as "undercover customers" and get paid for providing valuable customer service feedback to businesses. For cheaper products, they can even keep the product after the secret investigation or get it reduced in price, which is natural when the product is a service.

Thus, mystery shopper reports can be invaluable tools in measuring customer satisfaction and experience, along with salesmanship, professionalism, compliance, closing skills, and follow up. The reports should provide feedback in support of training, performance reviews and incentive programs. Most important, detailed feedback can help to resolve problems before customers or markets are lost. Additionally, it can help to recognize and reward "superstars" before losing them to mediocrity or to the competition.

Before an investigation is done by a mystery shopper, she/he is often trained to memorize a check list, which should be the minimum of information the company wants to have investigated. This is because being a mystery shopper means that she/he cannot use a written check list or a piece of paper to take notes on during the investigation as that would be a give-away to those being investigated. Therefore, so as not to forget any valuable information, a recommendation should be filled out as fast as possible after the investigation or simple notes made that can be elaborated on later. Using a recorder from which a report can be later typed is even better.

8.9.4 On-Site Shopping

The point-of-sale is the critical moment of truth in determining whether a customer is satisfied or not. Evaluation of the interaction between the mystery shopper and the front end people is therefore important in understanding the type of service the customer is receiving and whether sales associates or team members are following prescribed procedures and protocols, if they make use of opportunities, or if they are creative. In essence, on-site shopping gives management input to look at sales and service efforts from two perspectives—that of the customer or guest, and that of the organization. It does not matter if the evaluation is for a retail store, a restaurant, a car dealership, an apartment development, or an amusement park.

On-site shopping also has the advantage of the investigators being able to assess the physical aspects of the location including the cleanliness and upkeep of the facility, the proper use of merchandising and marketing materials, and the appearance of the staff and their attire (including uniforms), product presentation and demonstration.

8.9.5 Phone Shopping

Phone shops can be used to evaluate a number of different dimensions of the telephone transaction. Was the telephone answered promptly? Having the phone ring too many times, or placing the customer on hold for an extended time, is not a good way to start. Did the tone and manner of the associate reflect the spirit and attitude you are trying to project? Was the associate able to speak intelligibly and correctly answer the questions the customer had? Did the associate address them in a welcoming manner and invite them to visit? All of these elements can be evaluated through the use of quick, low-cost telephone mystery shops.

The best way to fully appreciate the telephone interaction between an associate and the customer is by recording the call as part of the telephone mystery shop. Recording of the phone shops provides a training vehicle to sit down with the associate to point out strong and weak areas. Note that the recordings will not, for ethical reasons, be made available to anyone other than the associate and her/his closest manager. Before phone shops are done, all associates should be informed that such investigations will take place, which can have a positive side effect in itself with regards the positive way potential customers will be handled.

8.9.6 Internet/Web Shopping

For an increasing number of industries, customer contact is initiated over the Internet or via the Web pages, Facebook, and Apps. The speed and quality of the response over the Internet or via the Web page is just as important as an interaction in person or over the phone. Internet/Web shopping is done to evaluate response time to inquiries or orders. However, not only is the timeline important, but the quality of the interaction as well. Potential customers want their questions answered. Sales managers want potential customers to be taken through the sales process effectively resulting in successful transactions and revenue.

A special interest should be to measure how efficient the Web pages are, how long it takes from entering the Web page until the required information is found, necessary actions are taken, and the customer leaves the Website.

Increasingly, the Web pages must be made to allow all types of visitors to access them including, for example, blind people, color blind people, and epileptics. By having representatives for different user categories test the Web pages as well as letting experts test them, “Web for All” can be achieved, thus increasing market potential (Björk et al. 2008).

Chapter 9
Sales of New Products

9.1 Introduction

Until a mature situation has been reached for an innovation on the market, the different marketing and sales activities should preferably be handled initially by the entrepreneur. Ideally the entrepreneur is in the center of the activities organized in a Planetary Organization as is shown in the Fig. 9.1 for sales activities. If she/he is successful, more specialized team members can be added as soon as the economic situation permits. Therefore, each small circle in the picture does not mean one unique team member other than for large and mature businesses. Initially, all activities are often done by the entrepreneur.

9.2 The Importance of the First Order

The most optimal situation for the development of an innovation is to sell a new product idea before it is a fully developed product. Such an order prevents the product developers from developing "wrong" products and/or to "over developing" them. This is because the developers have a customer and/or user to please, which will make them more focused compared to developing the product without an external customer and/or user.

The price set for the first customer of a new product can be heavily discounted, however not to the value zero, as the most important thing is to get the first order and not to make profit on the first customer. This is as customer commitment is extremely valuable for starting up the development of a new product. An agreement can also be reached with the customer that payment can be made if and when, after a test period, the product works to the customer's satisfaction. However, the real price should always be clearly stated independent of the discount given.

S. Ottosson, *Developing and Managing Innovation in a Fast Changing and Complex World*, https://doi.org/10.1007/978-3-319-94045-8_9

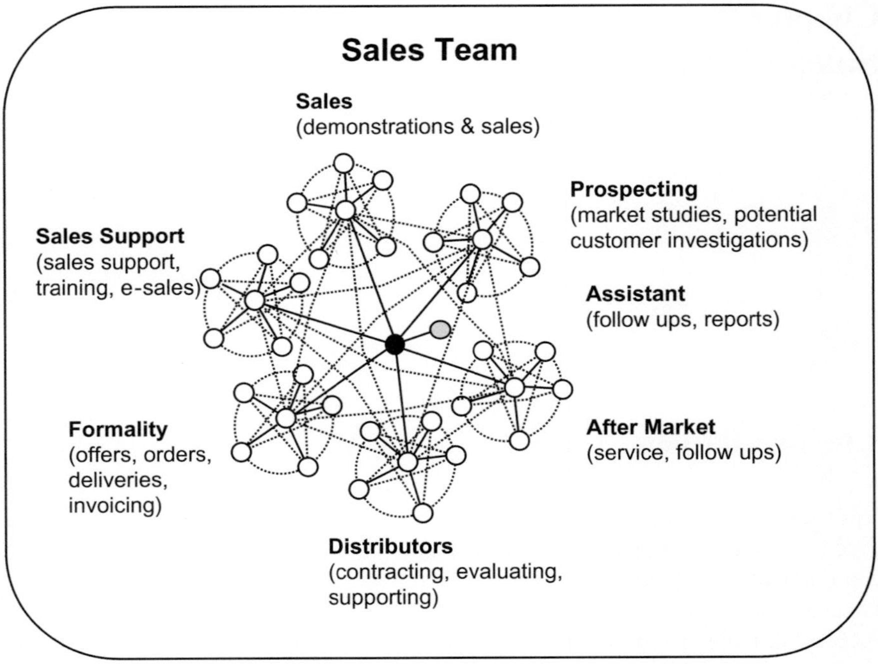

Fig. 9.1 Different sales activities to take care of when the innovation has matured to be a standard product in the company

The first order also has a great psychological effect on the salesman, the organization and the financers. The first order means that the salesperson has succeeded in convincing others of the benefits of the product and managed to develop a selling story, which in turn indicates that it should be possible to sell the product to other customers. These positive effects remain even if the product has been sold at a lower price than the costs were to produce it.

In order to get a complete picture of possibilities and problems with the product, the entrepreneur her/himself, ideally, should take care of the sales of the product idea from the start of an innovation project. By testing and refining the product/business idea on different people, met casually or at planned meetings, the product/business idea is gradually improved and models of the product/business idea can be developed to meet the often vague demands initially set. Built on the experiences from sales a selling/convincing message can gradually be built up.

If the product cannot be sold before it is fully developed, then, in many cases, the question should be asked if the development should start at all. For radical new products, this statement may be too far reaching but at least it must be possible to create a strong interest with one or more potential buyers.

Thus, an important principle to rapidly build a sustainable new business on its own or in a company is that the sales of a new product should come at least before the final product development is done.

9.3 The Art of Rhetoric

When a salesman addresses a potential buyer, she or he needs to use rhetoric as a tool to persuade a potential buyer to purchase the product she/he is offering (see next chapter). A short definition of the term *rhetoric* is that it is "the art of speaking or writing effectively" (Webster's Definition). According to the ancient Greek philosopher, Aristotle, rhetoric is "the ability, in each particular case, to see the available means of persuasion." He described three main forms of rhetoric as Ethos, Logos, and Pathos. According to Ramage and Bean (1998):

- *Ethos* (Greek for 'character') refers to the trustworthiness or credibility of the writer or speaker. Ethos is often conveyed through tone and style of the message and through the way the writer or speaker refers to differing views. It can also be affected by the writer's reputation as it exists independently from the message, his or her expertise in the field, his or her previous record or integrity, and so forth. The impact of ethos is often called the argument's 'ethical appeal' or the 'appeal from credibility.'
- *Logos* (Greek for 'word') refers to the internal consistency of the message, the clarity of the claim, the logic of its reasons, and the effectiveness of its supporting evidence. The impact of logos on an audience is sometimes called the argument's logical appeal.
- *Pathos* (Greek for 'suffering' or 'experience') is often associated with emotional appeal. But a better equivalent might be 'appeal to the audience's sympathies and imagination.' An appeal to pathos causes an audience not just to respond emotionally but to identify with the messenger's point of view—to feel what the messenger feels. In this sense, pathos evokes a meaning implicit in the verb 'to suffer'—to feel pain imaginatively. Perhaps the most common way of conveying a pathetic appeal is through narrative or story, which can turn the abstractions of logic into something palpable and present. The values, beliefs, and understandings of the messenger are implicit in the story and conveyed imaginatively to the reader. Pathos thus refers to both the emotional and the imaginative impact of the message on an audience, the power with which the messenger's message moves the audience to decision or action.

Thus, in short, a good story that is logical and told by an engaged and trustworthy representative will help to get an order.

9.4 The Initial Sales

To be successful with selling a product is more a question of selling a good story than hard-selling the product. We should also bear in mind the saying that "People love to buy, but hate being sold".

For the entrepreneur, or someone selling a new product, to be able to get the first order, she/he first has to learn about the product idea or the product and its background. To be successful she/he must be excited about it, make a convincing story, and then tell the story to other people she/he meets.

An example of testing a selling story is to start telling a little bit of the story, then shut up and wait for some comments or questions. If there is no such feed-back, tell a little more and shut up again. If there is no feed-back on this second try, either the person is not interested or the salesman is telling the wrong story. Therefore, modify the story and test it on the next person.

When the salesman, sooner or later, has got the first order, she/he knows what selling points were useful. However, for the next order, she/he often needs to adjust the story/message and add information. Having got a number of orders, it is time to start to make supporting marketing material especially on the Website. If you are the entrepreneur, do not be too discouraged in the early selling process. It is good piece of advice to practice your stories on a friend or relative before you test them on unknown people.

The first order is often the most difficult one to close, and the less developed the product is, the greater the achievement. At the same time, the prospects for the future of a new business venture are best if, at the concept stage, one succeeds in selling the product to a customer who preferably is also to be the user of the product. User influence is thereby automatically built into the product, and the more demanding this user is, the better it is for the development work. This is because a demanding user forces the development to be adapted to the product that has already been sold, which prevents over elaboration. This is on condition that the product developers are not isolated from the user(s) of the product, which unfortunately is not uncommon in industry.

When selling new products, the seller will often receive a "no" from a potential customer. This "no" contains information that the seller can use, partly to improve her/his argumentation on customer benefit, partly to determine to which buyer category the potential customer belongs—pioneer, early majority buyer, late majority buyer or latecomer. Below are a number of different "no's" and their possible meanings:

- No, I cannot afford it (really I want to buy, but have difficulty in financing)
- No, I cannot decide now (help me to gain support with internal marketing)
- No, the product has the wrong design/ weight/ performance/ form/ color (the price is right but adjustments are required)
- No, we are looking at other alternatives (they are arranging a situation containing several competitors)
- No, I do not have time (respect this and get back later)

- No, I do not know enough (produce a short basis for decision)
- No, I do not speak to salespersons (call in a product developer)
- No, I cannot make a decision (talk to the boss)
- No, there are cheaper alternatives (interested but wants to test the pricing, make another offer)
- No! (interpersonal friction).

So a "no" is often a conditional "yes" if the situation changes and the seller learns to interpret and respond correctly. In all selling, one must try to avoid receiving a definitive "no". To turn a definite "no" into a "yes" is prestigious and always takes a long time, as well as demanding patience and tact to achieve success.

Therefore, avoid driving matters to a definite "no", but break off the attempted sale and keep the dialogue going until the customer is ready to say "yes". Most need time to mull over a decision to buy. One should therefore never be too anxious to force a decision before the customer is ready for it. If the customer is forced into making a decision too early, the easiest decision for the customer is a definite "no". When there are purchase signals, it is time to act fast and have the order signed. Good sellers feel when the time is right.

9.5 Satisfying Expectations

The more conceptual the product has been at the selling point, the higher the expectations of the new product can be. Therefore, it is important in development to ensure, first of all, that the functional values will be at a good level, i.e., that functionality exceeds a minimum level. Thus, the product must correspond to at least what are called basic properties. These properties are often not mentioned by the customer, since they are considered to be self-evident.

The difficulties that lie ahead when selling a product at the concept stage are both of a technical and of a psychological nature. An early order gives knowledge about the user/usage, but also a psychological lift, which means that life seems easier to the entrepreneur, the team, the company and to other individuals involved in the business. The business has thereby acquired an initial commercial base, in the shape of at least one actual customer. It is important to point out that, in this context, for both ethical/moral and practical reasons, it is wrong to sell a product when one is not familiar with its technology.

The tactics of selling a product at the concept stage, i.e., before the product development has started or is fully developed, can seem bold to many. Nonetheless it is a fairly common method that is often used by small enterprises that lack the financial strength to develop the product first. In certain sectors, it is also usual to sell development assignments that one can use for the development of one's own products and which the customer then can benefit from in the form of royalties on units sold to other buyers. This approach is often used for military products and systems.

If the product cannot or should not be sold at the concept stage, the goal should be to sell it as soon as there is a functional prototype to show. This prevents the risk of too much technological development without customer/user contact. The worst situation arises if the sale of the product does not begin until it is fully developed and adapted for production.

Note also that experience says that a user or user group that tests the product and perhaps also takes part in development, but does not make any financial commitment, does not provide as good input values to the product developers as when the group or someone whom the group represents additionally has made some financial commitment. The explanation is that when it costs nothing, it is easy to have views that one then need not answer for when actually paying for the product.

The *basic properties* of a product are often referred to as expected requirements. When a product meets such tacit demands, it does not mean that the customer/user becomes more satisfied with the product or company. However, there will be great disappointment if the product does not have the basic properties. Therefore, it is important that the person who sells the product ensures that development takes place in such a way that the basic properties are secured. If this does not happen, then sales will be considerably more difficult, due to the negative publicity about the product that will spread.

The next level of demands with which the seller must ensure reasonable compliance is the *normal requirements*, since these correspond to the price that the customer has accepted for the product. Examples of normal requirements for a car, to take a product example everyone is familiar with, are acceleration performance, fuel consumption, quality and reliability.

The customer/user often expresses wants about the normal requirements, and the better the normal requirements meet expectations, the greater the customer's/user's satisfaction. At the same time, the normal requirements mean increased costs in proportion to how extreme they become. Therefore, the seller must also moderate the extent of the price-raising activities that often occur if the product developers and customers together are given carte blanche for interaction.

There is also a third level regarding expectations. This concerns *exciting requirements* and is a response to the customers'/users' tacit or unconscious wishes and demands. Unlike expected requirements, the exciting requirements are not taken for granted, but constitute positive and unexpected surprises for the customer when the product is delivered. These exciting values can be represented by simple and low-cost services and items such as, for example, the customer finding a bottle of wine on the back seat of a new car with a congratulatory message on the new purchase, the discovery that the fuel tank is full or that there is free recovery insurance. It is in the interests of the seller to ensure that some exciting requirements are included with the product upon delivery, since they contribute to the customer spreading favorable assessments, among the customer's circle of acquaintances.

9.6 Price-Setting New Products

Every entrepreneur and company leader must strive as quickly as possible to achieve an aggregate income that exceeds aggregate expenses in the business. If customers pay, this means a positive cash flow. With a positive cash flow, money is available for development, investment and accumulation of reserves for future economic strain or future needs. The cash flow is heavily dependent on the price-setting of the new products discussed earlier, and on related services, and ensures that the customers will pay.

A simple way of price-setting a groundbreaking new product is to gain an impression of the usefulness the product will have to the customer/user. Then through practical tests, one must investigate which price is the highest that can be set without making the product impossible to sell to the buyer category ‘pioneers’. This is done mainly through actual attempts to sell to some selected customers. If a customer buys directly, without discussing the price, then the price is probably too low. If the reaction is favorable to the product but not to the price, then this can always be adjusted downwards so that a price level acceptable to the customer is reached.

If it is difficult to sell the product at the set price, introductory offers can be made. An alternative is to reduce the price, clearly explaining the real price. Note, however, that the method of asking a possible customer to suggest a price seldom gives a correct price indication unless one has first indicated a price. This applies especially if the customer is hesitant about buying the product.

9.7 Enhancing Sales

We are constantly influenced, manipulated, and affected by people, places, information and situations. Influencing is about being able to move things forward, without pushing, forcing or telling others what to do. Manipulating means, for example, to arrange for speeding up a sales process. Excellent influencing skills require a healthy combination of interpersonal, communication, presentation and assertiveness techniques. Manipulation needs some strategic thinking.

Acceptable manipulation in the selling process is based on high personal ethical and moral standards. Some examples of manipulative items that can be used when selling products are (Cialdini 2000):

- *Reciprocation*: People tend to return a favor.
- *Commitment and Consistency*: If people commit, verbally or in writing, to an idea or goal, they are more likely to honor that commitment. Even if the original incentive or motivation is removed after they have already agreed, they will continue to honor the agreement. For example, in car sales, suddenly raising the price at the last moment can work because the buyer has already decided to buy.

- *Social Proof*: People will do things that they see other people doing. For example, in one experiment, one or more confederates would look up into the sky; bystanders would then look up into the sky to see what they were seeing. At one point this experiment was aborted, as so many people were looking up that they stopped traffic.
- *Authority*: People will tend to obey authority figures, even if they are asked to perform objectionable acts.
- *Liking*: People are easily persuaded by other people that they like. The marketing of Tupperware® in what might now be called viral marketing is an example of this, as people are more likely to buy if they like the person selling something to them.
- *Scarcity*: Perceived scarcity will generate demand. For example, saying offers are available for a "limited time only" encourages sales.

9.8 Customer Trust

Everyone who buys a product or a service normally has a trust that they get a problem-free product. If they gamble, buying something cheaper than what is a reasonable deal or at an auction, they must have trust in their own ability to get the product up to standard in some way. A customer without confidence in her/his own ability would prefer to buy a well-known brand with an extensive service network.

The buyer of a new product needs to feel confidence in the seller, since the product is unknown and maybe also the name of the company. Therefore, the customer's only security if the product does not function well is the seller's ability to rectify the problem.

In this respect, the situation is considerably different from sales of established products. With established products, the customer has a greater need to feel confidence in the manufacturing company than in the seller of the product. Needed confidence in the seller therefore decreases when the product becomes older and better established. Consequently, for many of our old standard products, very little in the way of active selling is required. In extreme cases, all that is required is receipt of orders, whereby there is little or no personal contact between the buyer and the seller.

Webshop sales have taken off considerably since 2017 because it is easy to find out different solutions "at home", to compare prices, and to get quick deliveries. Webshops offer a good possibility to globally market innovations also for new small companies also with the benefit that the customers often are less critical than when buying the traditional way.

9.9 Prospecting

To find customers for a new product produced by a new company is not an easy thing. One way, which the Internet supports, is to approach a large number of people telling them the good story about the product over and over again. From when the first customer has been confirmed, the key to increased success in sales is to ensure that those who already have the product are satisfied with it so that they will open doors to new contacts and prospects.

Looking at statistics and then making intelligent judgments of prospects to contact does, according to my experience, give much result and is, also according to my experience, often not successful although textbooks tell us that to find potential buyers one must first analyze the requirements set for buyers as regards financial resources, preferences, and geographical area. There is nothing wrong with that and having done this, one should keep one's eyes and ears open for new buyer categories. For example, good sources are local newspapers, trade journals, interest organizations, radio and TV news items.

Spontaneous contacts in connection with different events and meetings as well as at fairs and exhibitions are often good opportunities to create interest in a new product. Thus, new contacts taken face to face (F2F) with people wherever they happens to be—on the bus, in the cinema queue, at a club meeting, or at a conference,—can pay off in new customers. A large contact net will also help to increase the number of customers. To use persons who possess a large area of contact as door-openers is a well-known method to create business opportunities. To encourage them, door-openers may receive a fee on a commission basis on the sales made through their contact network.

Upon creating broader contacts, when one is confident of knowing which groups and persons should be contacted, then the telephone is a quick aid in setting up a contact and in sowing the seeds of interest in the new product. However, with an increasing number of companies using call centers, people get tired of telephone sales calls after working hours. Therefore, an e-mail contact or a SMS-message can sometimes be more successful as the first step in agreeing on a meeting. However, using the classical mailing method should not be forgotten.

For products of some direct or indirect value, showing the product improves the sales possibilities compared to telling about it or showing it on Web pages or in brochures. Therefore, it can be beneficial to try to get an opportunity to show the product to potential customers as the first step in selling it. Setting up a meeting time from a distance is mostly done by telephone or e-mail. However, sending e-mails to people not already known, generally gives very poor results. Telephone calls without referring to someone else, also, in general, show to give a poor result as many people are tired of telephone sales callers.

If there is reason to believe that it will be difficult to arrange a meeting via a phone call or e-mail, there is the alternative of going directly to the person in the hope of having a short meeting at once or later. This can occasionally be the only way to make contact with stressed-out managers. If such a possibility, by chance, appears, it is often when they are sitting on an airplane or a train and are approachable in a way that is not possible when they are involved in ordinary operations.

If possible, do not ask the person to come to the sales office, but go to the person instead. This avoids the cancellation of many meetings, at the same time, as the setting and environment around the buyer will help the seller to give appropriate information and use appropriate argumentation to plan the sales tactics as well as possible. An on-site visit means also less inconvenience for the potential buyer. The only thing risked by the person is the time taken to listen to the presentation.

If, on the other hand, the buyer is to come to the seller, the buyer will be considerably more inconvenienced, which reduces the likelihood of a meeting. In addition, this means that the seller misses valuable impressions gained from the buyer in her/his environment. To call on a person means, or at least is often perceived to mean, doing the person a favor, which gives a psychological advantage at the start of the conversation. It then becomes more difficult to dismiss the seller.

9.10 Sales Tip-Offs

There is abundant literature on sales techniques, sales psychology, and body language. Therefore, these subjects are not treated at length here. However, some important matters will be pointed out that apply specifically to sales of new products. When selling a new product, the task is to create confidence and, as far as possible, to avoid obstructing factors and effects. For example, to be interrupted by a mobile phone is perceived by most people as very irritating. Confidence is created, to a large extent, through language and regarding language "When in Rome, do as the Romans do" is an old but wise rule, which will make it easier for the seller to become accepted by her/his prospective customer. To use simple language reduces the risk of misunderstanding even if the person one meets has an extensive vocabulary.

The use of neutral and inconspicuous dress is recommended to avoid putting potential customers in a defensive mode that will need to be dealt with. A seller of new products, who travels around in expensive cars and otherwise is surrounded by expensive accessories, sends out signals that she/he maybe charges too much for the products. (This is, however, not the case for multi-level-marketing—MLM—distributors when recruiting other distributors to their network.) In many cultures, it is wrong to parade one's success in this way when marketing new products. In the same way, it is wrong to dress down too much when meeting persons with a lifestyle of a certain standard. Therefore, it is important to have salespersons that fit into the environment. For example, there is a big difference between selling motorbikes, pharmaceuticals and office equipment.

When a good two-way communication takes place between buyer and seller, the chances should increase of getting an order than if this situation does not exists. If communication is also good between other persons influencing buying, then the time to a purchase decision could be shortened. Therefore, one should strive to arrive at a situation as quickly as possible where the potential customer and other influencing persons become involved and actively participate in discussions about the product. This can be achieved by asking questions that invite involvement. The potential customer and others should also, if possible, hold the product in their hands and/or test it, so that they gain a feeling of involvement with the product.

Reaching a situation in which the potential buyer and other influencers become actively interested and involved in the product demands that the seller acts pedagogically and psychologically well. Expressions and body language need to be interpreted so that each personal meeting is adapted to give the best sales result. Stereotype and well-practiced sales conversations often become impersonal and uninspiring. Personally planned meetings without the use of too much ready sales material, therefore, often will provide better opportunities for contact and shared enthusiasm.

A seller who occasionally can remain silent forces the potential customer and other influencers to talk and, thereby, to reveal their opinions on the product. Silence, as a sales instrument, is often a highly effective tool to resolve a deadlock. This is because most people seem to shun silence; something that may be interesting to try out on one's own acquaintances on some occasion. If, in addition, it can be arranged for the prospective buyers to hold the product, it is practically inevitable that they will start speaking about it. Since it is unpleasant to say negative things about a new product, it comes naturally to most people to say something positive, which gives the seller an opportunity to reinforce these judgments. (This reasoning, from an ethical point of view, presupposes that the seller is convinced of the quality of the product and of its benefit to the customer/user.)

To be humble and considerate helps to prevent personal friction between the buyer and the salesman. A proverb says that honesty is the best policy, and it is generally better to admit that one does not know everything than to give a reply that perhaps is incorrect. In addition, unanswered questions give the seller an opportunity to get back to the buyer and to keep up the dialogue until a purchase takes place.

At all meetings with potential customers, it is important to achieve a feeling of personal community, which leads to confidence and a possible purchase of the product. Each seller should, therefore, as discretely as possible work as quickly as possible through the wall of reservation and reticence that most people have when encountering an unknown person. If, on top of this, the stranger has come in order to sell a product that the person has not planned to buy, the defensive instinct will be stronger than normal. For this reason, it is advantageous not to begin sales efforts with completely unknown individuals, but with someone with whom one has already built up some confidence. To be recommended by someone who knows the individual is often considerably more favorable than contacting a completely unknown person oneself.

In all pedagogical activities, repetition is the "mother of knowledge". Since it is of the utmost importance that the buyer should learn to understand and have confidence in the product, it is important to make diligent use of repetition. Talk about the product and demonstrate it in different ways, so that the repetition is not felt tiresome. Use questions to check that the buyer has understood the message. Dialogue is more important than discussion when selling a new product. Once the individual has bought the product, it is time to get her/him to feel proud of it without feeling abandoned by the seller. Consequently, stay in touch with the buyer and quickly deal with any faults and misunderstandings. This will ensure that a valuable reference person and ambassador for the product have been gained. In this person's circle of acquaintances, there are probably new potential customers who can be approached. In time, the buyer will also be ready to replace the first product with a new, improved or further evolved product, which is a reason to take good care of this customer. A disappointed customer tells many of her/his experiences, while a satisfied customer unfortunately seldom mentions the fact to anyone.

One trap that many sellers fall into is to speak badly or disparagingly about a competitor or a rival product. This always backfires sooner or later. One's own product is supposed to be so superior that one does not need to argue in a negative manner. Sales based on the seller's success in proving how bad another product is are not particularly useful for successful sales. In addition, a customer who has bought a product for negative reasons is hardly a good ambassador for the new product.

Another trap for sellers of new products is to focus on prices, as the buyer of a new product should not buy the product primarily because it is cheap. To focus on the price, therefore, is to stress the wrong aspect. The focus must instead be on the customer benefit of the product and the total offer including installation, discounts, and payment details. If these are accepted by the buyer, the price will often be a minor issue.

Many sellers give away products and promotional material. To give away products is a questionable marketing method, since the recipient does nothing in return. Gifts of this type are usually poorly utilized since there is little commitment from the recipient. Different types of promotional material that have nothing to do with the product hardly provide an incentive for the customer to buy the product. However, such gestures do reinforce the image value of the company and its products.

To summarize, one cannot overemphasize the importance that the seller acts as a fellow human being and helps the customer when necessary. A seller who lacks the ability to understand a customer's problems, preferences, needs, wants, and wishes will find it difficult to sell a new product. The term "salesman/ saleswoman" can be an unsuitable one to use in this context. To use the term "account manager" on one's business card can also be unsuitable. A better choice than the title of "salesman/woman" on a business card would be "applications engineer", "service technician", or simply "representative".

9.11 Easily Made Mistakes

To build a loyal customer base demands skilled handling, takes a long time and is laborious. The work can quickly be undone if the customers and users are not cared for in an exemplary way. This loss means that one also loses the persons—the ambassadors—who support the business, for example, with free advertising in the form of positive judgments and recommendations. If one does not tend the customer base that one has, then one is forced continuously to build up new ones, which costs perhaps ten times more in all respects than properly tending one's existing customer base.

A common mistake made by sellers is to promise more than they can deliver. This creates unnecessary irritation among customers. Confidence in the seller and the seller's company is thereby eroded. It is always better to promise less and deliver more, which leads to positive customer reactions to the seller and the company. Even if, in one's eagerness to sell, one finds it easy to make grand promises, in practice it is a virtue to be as modest as possible in making promises. This applies especially if the seller is not certain of being able to keep the promises. So, have generous safety margins in every respect when making promises to customers.

To prevent misunderstandings and unpleasantness upon or after delivery, take along the person or persons developing/adapting the product to visit the customer and most of all, the user or use environment. Check also during this work that the product really turns out as promised. User involvement is recommended in the product development process.

Another mistake, and an unethical one, is to sell too many units of the product to a customer or to sell products without knowing whether or not they will bring benefit to that customer. Fairly soon, the customer will discover that she/he has been manipulated in a way that will not improve her/his opinion of either the product, the seller or the company.

If the wrong product has been bought unintentionally or for some other reason, the company should rectify the matter as soon as possible. If the buyer is a retailer, it is important, in the same way, to support her/him in different ways, so that the products do not "die on the shelf". This will ensure an eventual repeat purchase.

9.12 Field Work

A big marketing drive in the mass media, in the same way as a mass send-out to selected categories of people, generally produces meagre sales results. Regarding direct advertising, when, for example, addresses have been bought from an address register, orders usually come from somewhere between one per thousand and a few per cent of the recipients of the direct advertising, which means that sales costs are high when postal services are used for distributing the advertising.

When the Internet is used for direct advertising, then the response frequency is of minor importance, since the costs are limited to the time it takes to search for e-mail addresses and to enter them in the address frame. Advertisement on Facebook follows other rules that will change from time to time.

Thus, it can be stated that direct advertising, whether distributed by post or Internet in different forms, generally gives a meagre sales result. Really good sales results for a new product demands effective selling in the field by salespersons who can inspire confidence in themselves and who can utilize market rhetoric in a good way by focusing on customer benefits and selling a solution instead of a product.

In a similar way, advertising campaigns often produce no more than an insignificant lift in sales of new products, except possibly in the case of fashion-oriented consumer products. To produce costly advertising material and to run broad campaigns when the product is in an early stage of production is often an uneconomical way to use money. Naturally it is nice for the sellers to have at their disposal brochures in four-color print, professional videos or USB memories with the information, and to be able to refer to an advertising campaign.

However, since sales are so strongly centered on the salesperson in the first stages, personal selling is often quite sufficient, along with a Web or Facebook page or a Youtube video with additional information. To raise the re-collection value, it is also recommend to have a business card bearing a personal photo.

Considerable capital is usually invested in the introduction of new products via trade fairs. Fairs are certainly visited, above all by people belonging to the categories "pioneer buyers" and "early majority buyers", which means that one can make many positive contacts. However, results in the form of sold units are often poor from a short-time perspective. One must usually settle for positive judgments, which must be taken for what they are. Even if all the visitors who visit a stand talk favorably about a product, there are very few who place an order directly during or after the fair. As always, it costs nothing to be favorable towards a product if there is no need to buy it. To achieve a good result from a fair, the personnel must, therefore, actively make follow-up contacts and influence potential customers after the fair. A positive side effect of fairs is that one will meet people who one would never have met otherwise, which eventually can lead to profitable outcomes.

As cannot be too much stressed, to be able to sell a new product successfully demands, in general, that the seller has direct contact with selected potential customers. In the beginning, if the entrepreneur alone runs sales, it is important that the entrepreneur spends a large part of her/his time in the field selling/marketing the products. If this is not done, then the entrepreneur has little chance to start up the operation, just as she/he does not acquire the genuine understanding of the market that is required for her/him to make good decisions, or to make demands on the sellers that are subsequently appointed. When one or more salespersons have been appointed, their work duties must give an even higher priority to sales in the field in comparison to in-house work.

If sellers of mature products are commissioned to sell a new product, the experience is often a shock to those who are forced to acknowledge that the telephone is not ringing or customer e-mails are not coming. They themselves must constantly make calls and actively arrange meetings and "sell themselves" to increase sales of the product. Selling new products is highly strenuous and can be a mental strain on a salesperson, since extremely close contact with the customers is needed. If problems occur and the worst comes to the worst, the customers can call the seller at any time of day, irrespective of holidays, to obtain a fast solution to the problem that has arisen.

The fact that early sales often take place through close personal contacts means that a seller cannot make that many sales calls per day. Even so, the seller should be disciplined and be sure to make a certain number of calls every day. Using modern technology and constantly updated financial data on the approached companies via the Internet, one can avoid unnecessary customer losses. With such information one can also set up suitable visiting rounds to potential buyers.

A sale is very rarely closed at the first meeting. If the meeting leads to the opportunity to make a written offer, that is to be considered as a good outcome. Before an order is placed, prompting and follow-ups are often required a number of times. These promptings and follow-ups can be done by phone and possibly also by e-mail, SMS messages and even, although seldom, by mail. It is not uncommon that the time it takes from first contact to order can be years. Therefore, "hard selling" to get an order at the first meeting is questionable. Maintaining a contact by helping the customer to make the final decision step by step is often a better method, especially for selling new products and services.

When the product has got some foothold on the local market, one should be able to achieve an outcome of 50–60% of orders placed in relation to written offers made. Note however, that in this case, the time from an offer to an order can also be very long and that many refined time-limited offers can be needed before the order has been received.

To achieve good sales results, it helps if the seller has a range of products to offer as a complement to the main product.

In sales contacts, the sellers will receive many good hints as to what can be done with the product to improve it and/ or new applications for the product. It is extremely important that this information and the opinions are promptly fed back to the company, so that the company has the possibility to utilize them in refining the product or in developing new products. One major advantage of a small organization is the ability to quickly capitalize on an opportunity that arises, while a large and unwieldy organization often "investigates to death" such opportunities.

9.13 A Final Remark

To be a successful salesman of new products and services is hardly possible without being good at networking. Therefore, be active in different networks and clubs so that you can use the contacts different persons have to get access to their contact net. By using the contact net persons A–D in Fig. 9.2 you get a contact of 35 or more persons to use. For example, having a Rotary pin on your blazer means that

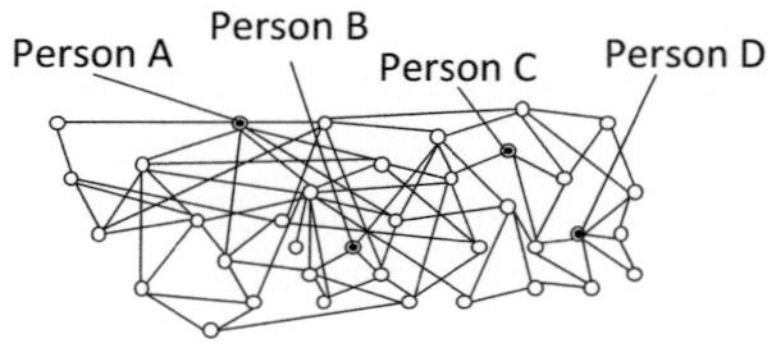

Fig. 9.2 By using other people´s network, your own network expands

you can skip all introductions when you meet another Rotarian anywhere in the world. (There are more than 32,000 Rotary individual clubs operating in over 200 countries. The members of all these autonomous clubs are called Rotarians, and together they form a global network of 1.2 million business and professional leaders, all volunteering their time and talents to serve their communities and the world. Individual Rotary clubs belong, in turn, to the global association called Rotary International.)

Chapter 10
Financing Innovation

10.1 Introduction

For an innovation to appear, there must be a (business) idea that an entrepreneur takes care of. In turn, for a good result, the entrepreneur needs a good enough environment around her/him. A fourth important component is capital, which is the focus of this chapter. Figure 10.1 shows pre-requisites in a principal way for an innovation to appear.

Especially in the start-up phase of an innovation project, the importance of the entrepreneur is dominant (Fig. 10.2). Therefore, the entrepreneur will be the first to be engaged in a new innovation project. She/he then chooses the team members to be enrolled in the project.

Before the incomes from the activities exceed the spendings—until "break-even" has been reached—external (risk) capital is needed. However, the return on the investments done will be negative until at least break-even is reached (Fig. 10.3). Note that the "J-curve" also reflects the cash flow for new businesses although it never will be as smooth as shown in the Figure.

Figure 10.4 shows an example of three real curves from a start-up enterprise during its first twelve months.

There are some ways to influence the cash flow in a positive way, reducing the need for external money. Six examples are:

1. Borrow equipment
2. Buy used equipment
3. Lease equipment
4. Work on payment terms
5. Increase selling prices and shorten payment terms
6. Sell invoices.

Another way to minimize the need of external money is not to start the development with a challenge but with new knowledge as Fig. 10.5 shows (c.f. Fig. 2.1).

S. Ottosson, *Developing and Managing Innovation in a Fast Changing and Complex World*, https://doi.org/10.1007/978-3-319-94045-8_10

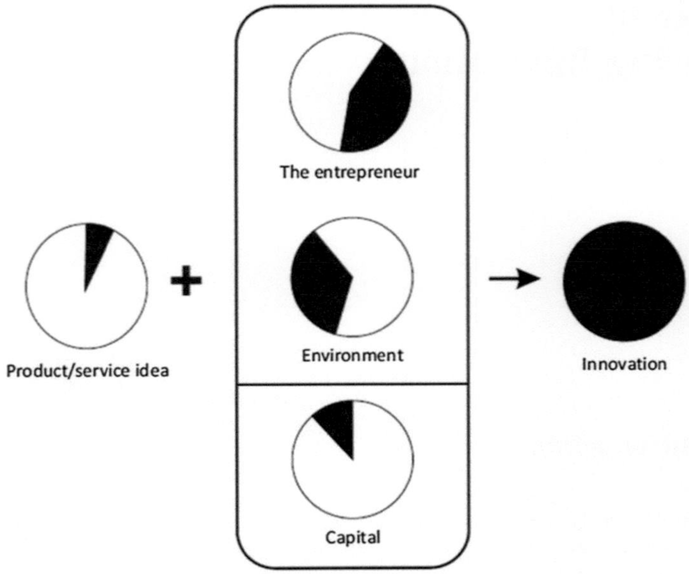

Fig. 10.1 Four important components to make an innovation

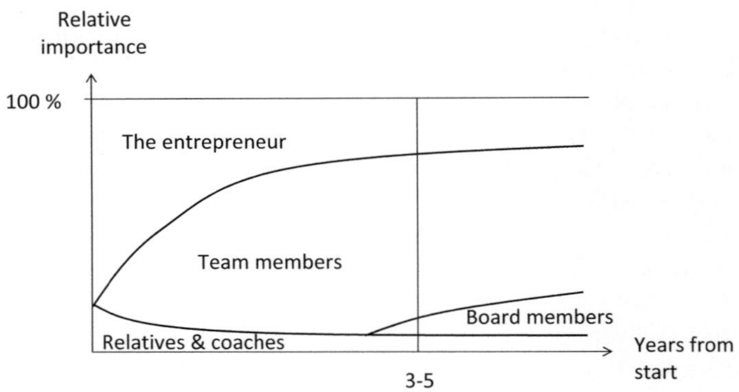

Fig. 10.2 From a successful start, the incoming staff will play an ever increasing role to establish the innovation on the market

If one or a company decides to start the development with a wish, the risk of failure can be minimized by choosing an entrepreneur high up in the stairway shown in Fig. 10.6.

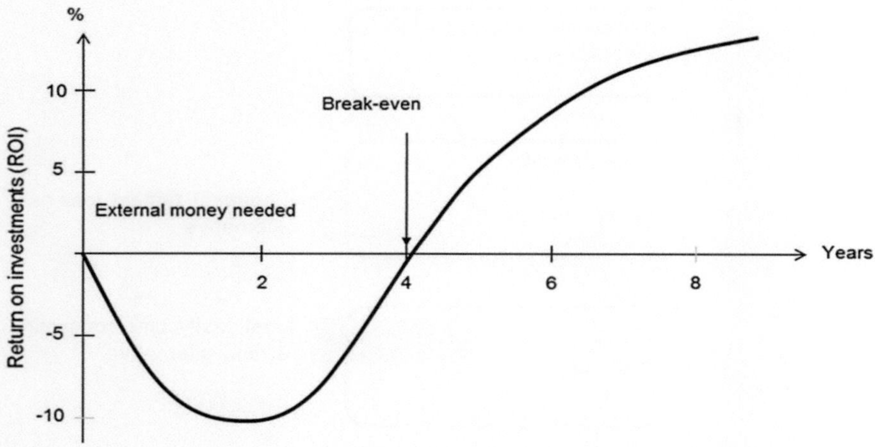

Fig. 10.3 The J-curve counts in general as well for ROI as for the cash flow of a new business

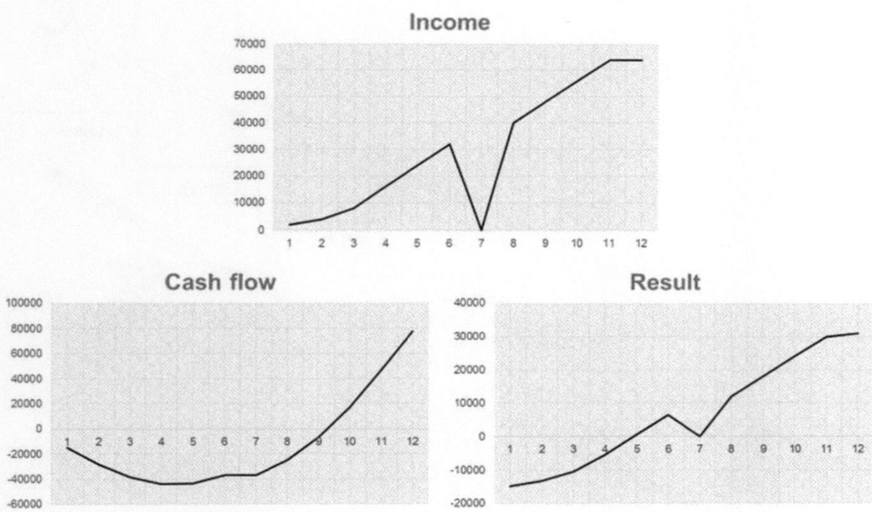

Fig. 10.4 Three important curves for a start-up enterprise during its first twelve months from the start

10.2 Risk Capital

Money that is invested in a new business is called risk capital as the investor must calculate that she/he will not get it back if the sales do not take off and the margins are not satisfactory. The most common risk capital is the money the starter of a company uses to buy shares for in the new company.

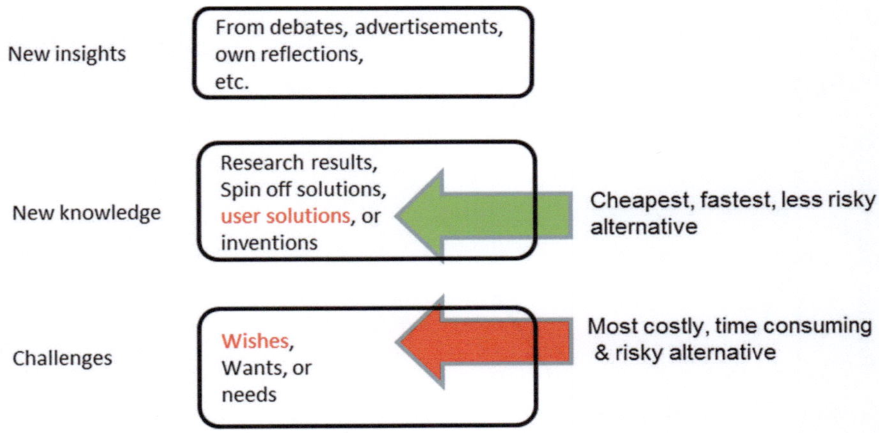

Fig. 10.5 There is often a choice from where to start the development of a new business

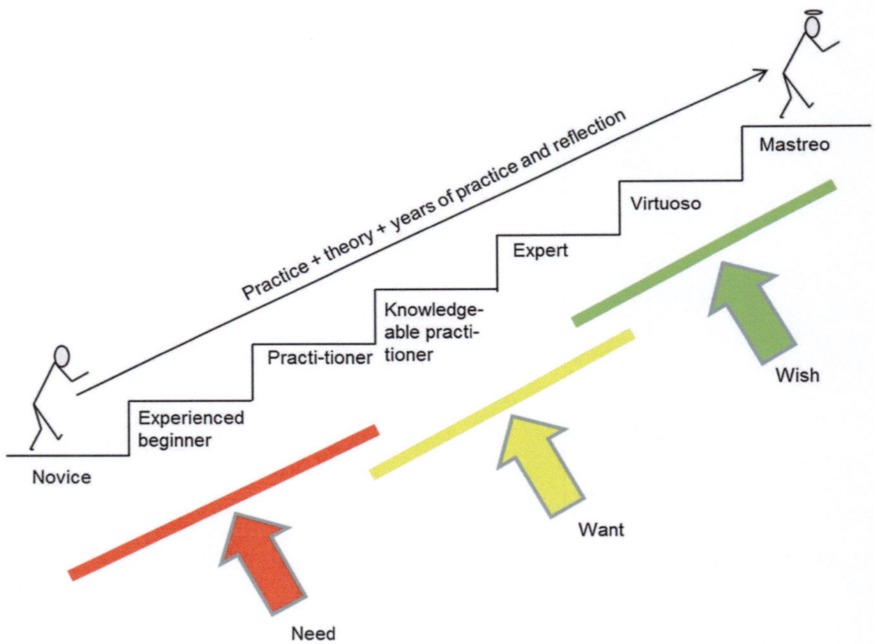

Fig. 10.6 If the start is from a challenge in Fig. 10.5, a suitable entrepreneur needs to have different knowledge and experience

For the financing of a growing business, there will be a need for monetary financing steps until there is a mature business. These are some common steps in the funding of a new and growing business:

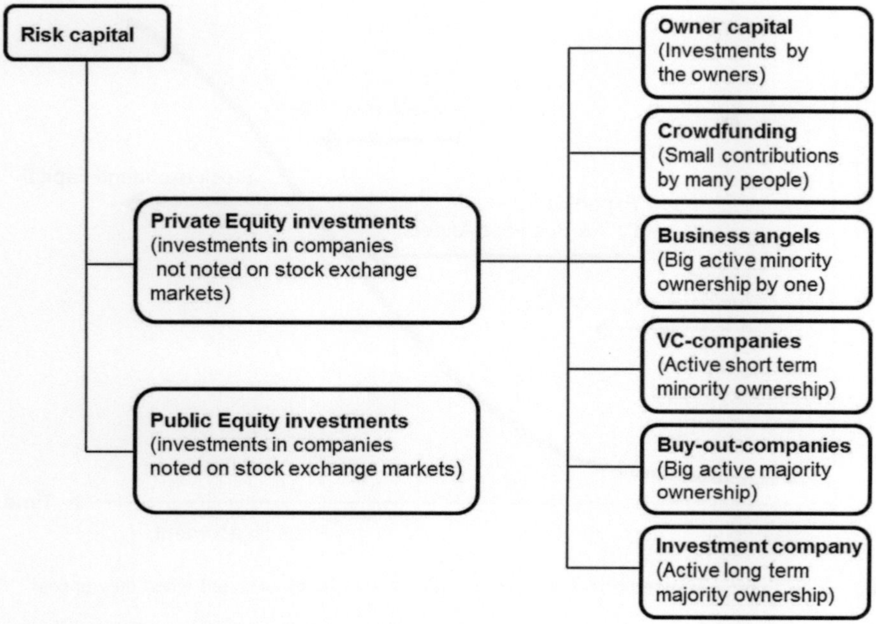

Fig. 10.7 Some common risk capital sources

1. If the new product is developed in a new startup firm, the first round of funding is often done by one or more starters.
2. For the development, sometimes crowdfunding can be used to get development money (Crowdfunding means raising capital, online or otherwise, often as small contributions from large groups of people).
3. For expansion, an alternative is to turn to business to so-called angels for money coupled, with their experiences and engagement. Business angels are people that generally have made money on their own by bringing up new businesses.
4. At a later stage, funds can be made available for small businesses with growth potential by Venture Capital (VC) firms. They provide managerial and technical expertise.

Figure 10.7 shows these and some other risk capital possibilities.

Figure 10.8 shows when different risk capital sources are active after the initial start of a new business. Money invested early after the start is sometimes called Seed money.

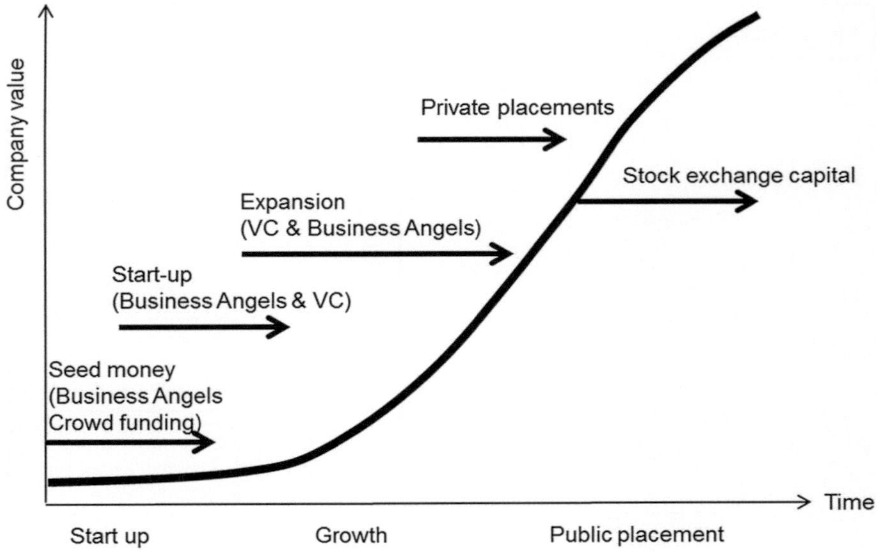

Fig. 10.8 Some common terms connected to risk capital investments and when they appear

10.3 Venture Capital

Venture capital (VC) is financial capital mostly provided to early-stage, high-potential, high risk, growth startup companies. A venture capital fund makes money from the sales of equity (the exit) in the companies it has invested in up to ten years earlier. The typical venture capital investment occurs after the seed funding is round (see Fig. 10.7).

In addition to business angels investing and other seed funding options, venture capital is attractive for new companies with limited operating history that are too small to raise capital in the public markets and have not reached the point where they are able to secure a bank loan or complete a debt offering. In exchange for the high risk that venture capitalists assume by investing in smaller and less mature companies, venture capitalists usually get significant control over company decisions, in addition to a significant portion of the company's ownership. Although venture capital is often most closely associated with fast-growing technology, healthcare and biotechnology fields, venture funding has also been used for other more traditional businesses.

The basic principle behind the VC shown in Fig. 10.9 is that many limited partners invest money in a VC fund managed by a VC company. The VC fund invests in many portfolio companies together with other VC funds. For each portfolio company, one of the co-investing VC companies is appointed to work closely together with the company to make it grow. That VC Company gets a

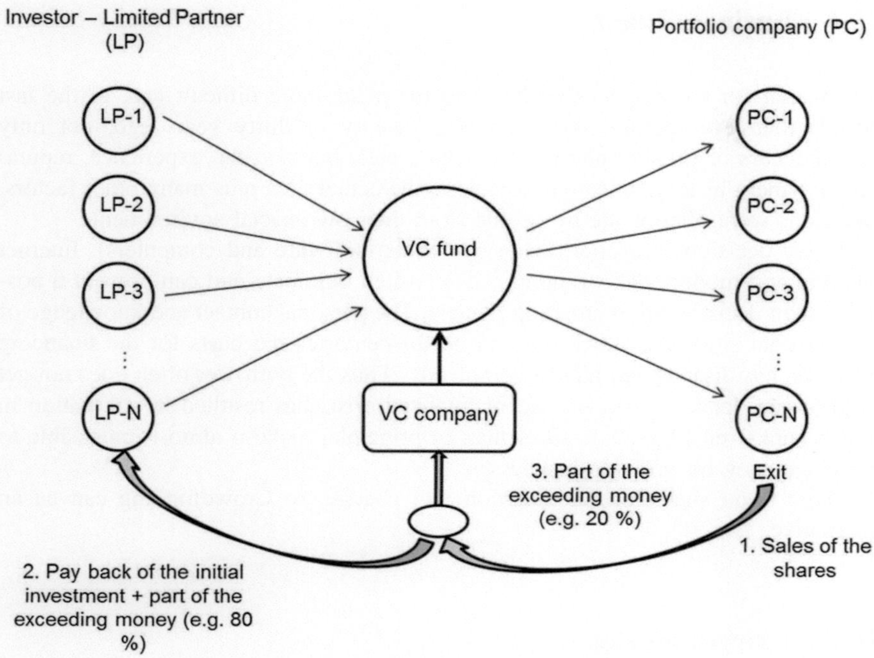

Fig. 10.9 The organization of VC

smaller yearly management fee (e.g., 2% of the investment) for the work and responsibility.

About ten years after the initial engagements in the portfolio companies, the shares are sold. When all the costs for the sales are paid, the initial investment is re-paid to the investors. What then remains is shared 20% for the VC Company and 80% to the investors.

American VC is based on the following outcome after about ten years:

> 10 investments will give:
>
> - 1 high return of investments
> - 3 medium returns on investments
> - 3 bankruptcies
>
> Wished return on investments: 20–30

10.4 Business Loan

The search for working capital has become a far more difficult task in the last decade than ever before. As recently as twenty or thirty years ago, not only entrepreneurs or private individual played a role, but also the experience, reputation, payment history, social environment, the actual idea plus many other factors. Decisions were often made by people from their own social environment.

Today decisions are often made by robots (software and computers). Internet banking and modern data communications often dominate and can prevent a positive credit decision on the funding request. The personal contact and knowledge of the personal situation, which used to be for centuries the basis for the financing business, has disappeared almost completely. Thus the borrower often does not get to know the lender. Also, the recent financial crisis has resulted in a situation in which banks must have securities that, in principle, make it almost impossible to borrow money for growing businesses.

There is no sign that this situation will change, so Crowdfunding can be an alternative.

10.5 Crowdfunding

Crowdfunding is a collective effort by people who network and pool small to medium-size investments together, usually via the Internet, in order to invest in and support efforts initiated by other people or organizations (Ordanini 2009). By 2025, the global crowdfunding market could reach between 90 and 96 billion USD according to a 2013 study commissioned by the World Bank (Noyes 2014).

Crowdfunding has three types of actors:

- the project initiator who proposes the idea and/or project to be funded,
- individuals or groups who support the idea, and
- a moderating organization (the platform provider) that brings the parties together to launch the idea.

In principle there are three principally different reasons why people invest in crowdfunding:

- donation to a good purpose,
- another way of purchasing a product or service, and
- equity or investment crowdfunding to gain ownership, or a promise of future returns.

Here is an example of a successful crowdfunding activity from Indiegogo (www.indiegogo.com 2016):

Fig. 10.10 The inventors Stuart and Cedar Anderson with their Honey-On-Tap Beehive (www.youtube.com/watch?v=0_pj4cz2VJM)

"Extracting honey from hives is arduous and time-consuming for beekeepers and stressful for bees. Australian father-son beekeepers Stuart and Cedar Anderson (see Fig. 10.10) set out the revolutionize beekeeping and honey harvesting, as well as to make it easier for anyone to keep bees. Their system allows beekeepers to harvest honey straight from the hive, without opening it. By turning a lever, the honeycomb cells inside the hive split vertically, and unfiltered honey pours out the tap. The bees are left relatively undisturbed and this also drastically reduces harvesting labor."

In late February 2015, the crowdfunding campaign of the Honey–On–Tap Beehive surpassed its 70.000 USD goal within eight minutes of launching, raising 2.18 MUSD in the first day alone (www.indiegogo.com). In one month, more than 25.000 backers had raised 8.9 MUSD (www.popularmechanics.com 9th of April 2015).

Seen from the entrepreneurial side the Crowdfunding Centre's May 2014 report identified the existence of two primary types of crowdfunding (Clifford 2014):

- Reward-based crowdfunding: entrepreneurs pre-sell a product or service to launch a business concept without incurring debt or sacrificing equity/shares.
- Equity-based crowdfunding: the backer receives unlisted shares of a company, usually in its early stages, in exchange for the money pledged. The company's success is determined by how successfully it can demonstrate its viability.

Crowdfunding platforms make it possible to test a product idea first with potential customers before securing any funding. If there is a high demand for the product, it may be worth developing it. Here are some other advantages to utilize crowdfunding:

- It removes risk for entrepreneurs who do not want to have to take out a bank loan.
- In the event that the idea becomes popular, one can leverage the initial investment into acquiring some private investors who can contribute even more.

Crowdfunding can raise awareness, and initiate word of mouth. If the product becomes popular, the act of securing financing also becomes its own sort of grassroots PR campaign.

Appendix A
Two Views of Life

The development of products, services, and processes can be carried out as two extremes, and can be done based on classical development principles or dynamic principles. Using dynamic development principles has shown to cut down the development time and the development costs at the same time, and the user satisfaction and pleasure increases (Björk 2009b).

For organizational purposes, two opposite management strategies can also be used for the development—the classic or the dynamic. The classic strategy works well for a slowly changing and predictable world. The dynamic strategy is well suited for rapidly changing situations and for innovative activities.

The classic paradigm builds on thinking from the classical (Newtonian) mechanics, while the dynamic strategy builds on views from quantum physics (some people say quantum mechanics, which is less good, as the word mechanics comes from the old view of everything being like a machine). A summary of the both paradigms is shown in Fig. A.1.

Some important comparisons between each circle in Figure 0-7 are shown in the three Tables A.1, A.2 and A.3.

A.1 The Complex Versus the Classic View

The use of the term "complex" is often confused with the term "complicated". To understand the differences, it is best first to examine the roots of the two words. "Complicated" uses the Latin ending "plic", which means, "to fold", while "complex" uses the "plex" that means, "to weave." Thus, a complicated structure is one that is folded with hidden facets and stuffed into a smaller space. On the other hand, a complex structure uses interwoven components that introduce mutual dependencies and produce more than a sum of the parts. This means that complex is the opposite of independent, while complicated is the opposite of simple.

S. Ottosson, *Developing and Managing Innovation in a Fast Changing and Complex World*, https://doi.org/10.1007/978-3-319-94045-8

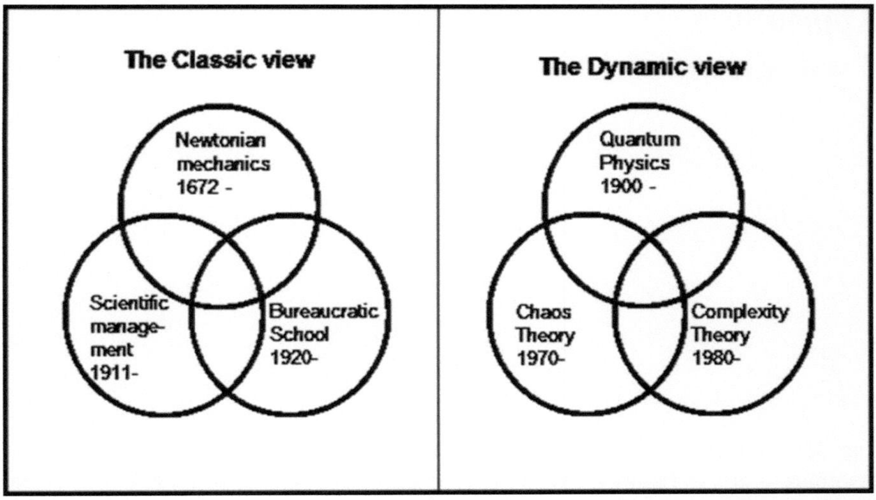

Fig. A.1 Some differences between the Dynamic paradigm and the Classic paradigm

Table A.1 Some important differences in characteristics between the classic and complex view

The Classic view based on the bureaucratic view	The dynamic view based on complexity theory
Closed system	Open system
Organizations can be controlled regulated	Organizations can only be influenced
Cause and effect is detectable	Cause and effect is difficult to explain
The whole is built up by elements	Elements are ignorant of the whole
Limited interaction in the system	Rich interaction in the system
Governing give expected results	Governing can give unexpected results
Slow feed-back from management	Short range interaction and feed-back loops on several levels
The reaction of the system is proportional to the impact	The system might react proportional as well as disproportionately to both small and large changes
One dimensional up-down character (line organization with an order of importance)	Network character (planetary organization with more equality among its members)
Behavior fostered by rules set by the management	Behavior develops from interactions within the system
Re-organization can take place after external agent interventions	Regular internal re-organizations without external agent interventions
Human emotions in the system often are not taken into account in the decision making at a higher level in the organization	Humans are affected by emotions why organizations operate far from equilibrium and have a history influencing the work

Table A.2 Some important differences in characteristics between the classic (Newtonian) paradigm and the quantum view

The Newtonian view	The dynamic view based on quantum theory
Reductionistic view (i.e. fragmentation, reduction, and isolation). Clear interfaces between each part in the system exist. Each part does not change much with time. Totality equals the sum of the pieces	Holistic view (i.e. synergy, interaction, and integration). Fuzzy interfaces between some parts in the system. Some parts can change drastically within a short time. Totality is more—or less—than the sum of the pieces
Only closed systems and hard objects/points can be treated on which different forces act (e.g. is $F = m * a$)	Focus on open systems and relations on which different fields act (e.g. gravitation, information, electromagnetic fields, etc.)
Only reversible processes can be treated	Also irreversible processes are treated
Small changes are negligible	Small changes can cause big differences
Periodic	Aperiodic
Deterministic/predictable—ultimately controllable	Indeterminate/statistic—uncertain fuzzy
The more detailed (microscopic) the measurements/studies are the higher the predictability will be	The more detailed (microscopic) the measurements/studies are the more uncertain the predictability will be (the Heisenberg uncertainty principle)
Linear—one best solution exists	Non-linear—many equally good solutions exist
Either/or	Both/and—superposition of states/amplitudes
Due to a so-called positivistic view, objectivity exists why the observer should not influence. Performing research or making market investigations should e.g. be done with an outsider/observing view	Even situation is unique (at least time changes). The observer always influences what is measured which is why the classical meaning of good research is difficult. In the open system die researcher is inside the system

To take care of the possibilities that appear in the un-linear world we live in, the use of the Complex Adaptive System (CAS) theory, which is a special case of complex systems, has shown to be of great value. This is because this theory seeks to answer some fundamental questions about living, adaptable, and changeable systems. (Examples of complex adaptive systems include the stock market, social insect and ant colonies, the biosphere and the ecosystem, the brain and the immune system, the cell and the developing embryo, manufacturing businesses and any human social group-based endeavor in a cultural and social system such as political parties or communities. There are close relationships between the field of CAS and artificial life. In both areas, the principles of emergence and self-organization are very important.)

Some important differences between the hierarchical/ bureaucratic view of traditional management philosophies and the complex adaptive system philosophy are given Table A.1.

Table A.3 Some important differences in characteristics between the classic (Newtonian) view and the chaos view

The classic view based on scientific management	The dynamic view based on chaos theories
A small change in the initial conditions will not change the long-term behavior of a system. A small amount of difference in a measurement is often considered experimental noise, background noise, or an inaccuracy of the equipment that is not important for the overall view	Dependent on initial conditions. Just a small change in the initial conditions can drastically change the long-term behavior of a system. Such a small amount of difference in a measurement might be considered experimental noise, background noise, or an inaccuracy of the equipment. Such things are impossible to avoid in even the most isolated laboratory
Some nonlinear systems are exactly solvable or integrable	Some nonlinear systems have no simple or closed form solution
It is possible to make accurate long-term predictions about the behavior of the system	It is not possible to make accurate long-term predictions about the behavior of the system
The more detailed (microscopic) the measurements/studies are the higher the predictability will be	The more detailed (microscopic) the measurements/studies are the more uncertain the predictability will be
Making long-term predictions is possible if all parameters are known	Making long-term predictions to any degree of precision at all would require giving the initial conditions to infinite precision
Chaos is destructive	Chaos is the ground for development. Self-organization occurs out of chaos and disorder
A system is either stable or unstable	A system can swing between chaos and order, chaos and order

A.2 The Quantum Versus the Classic View

In Quantum physics, the physicists study the behavior of matter and energy at the molecular, atomic, nuclear, and even smaller microscopic levels. In the quantum theory, energy is not continuous, but appears in the form of quanta (packages). It gives the underlying mathematical framework of many fields of physics and chemistry, including condensed matter physics, solid-state physics, atomic physics, molecular physics, computational chemistry, quantum chemistry, particle physics, and nuclear physics. The mathematics of quantum physics is advanced and it can be difficult to understand the world of quantum physics and phenomena in it (the book Alice in the Quantum Land (Gilmore 1995) can help to get some insight on a macroscopic level). However, from quantum physics, we have got useful philosophies/views also for product and business development. Table A.2 shows some differences between the classic view and the quantum view in our daily lives.

A.3 The Chaos Versus the Classic View

The Chaos theory attempts to explain the fact that complex and unpredictable results can and will occur in systems that are sensitive to their initial conditions. Table A.3 shows some differences of this philosophical approach compared to the classic view.

A used example of the chaos theory is known as the "Butterfly effect". This is because, in theory, the flutter of a butterfly's wings in one part of the world could initiate a weather development that ends up in a hurricane in another. In other words, it is possible that a very small occurrence can produce unpredictable and sometimes drastic results. This is because the development can be exponential from the start until a "critical mass" has been reached. When a critical mass has developed, the development is hard or impossible to stop. Also, from the chaos theories, we have got useful views for product and business development.

Appendix B
A Renovation Project

Paper presented at The International Conference on Business Case Research (ICBCR 2016) December 12–14, 2016, Hong Kong

Experiences with Planetary Organization

Stig Ottosson, Professor (Norwegian University of Science and Technology), Norway
Evastina Björk, Ass. Professor (Norwegian University of Science and Technology), Norway

Abstract

The use of traditional management principles using line or matrix organization have been shown to have shortcomings in organizing and carrying out complex and dynamic activities. Therefore, other organizational models and management principles are tested with a shift from managerial behavior to management technology, using so-called circular management practices, also known, for example, as holacracy and sociocracy. A third model, that was originally developed for new product development projects is called Planetary Organization, and combines dynamic management principles based on philosophies from quantum physics, complexity theory and chaos theory. It has shown interesting results. However, this third alternative, using a different philosophical view is also suitable for critical situations. This report is about the turnaround of a company in a critical situation, for which this third alternative was used.

The paper describes the set-up of the Planetary Organization, some comments on how the leadership and management were carried out on a daily basis, the result of the work, and ends with reflections/conclusions. The change management/ turnaround project started with the reshaping of the company to be a planet, with the sun in the middle (the author/CEO), moons (the departments) in orbits around

S. Ottosson, *Developing and Managing Innovation in a Fast Changing and Complex World*, https://doi.org/10.1007/978-3-319-94045-8

the planet, sub-moons (the groups) orbiting around each moon, and comets (the free specialists) who moved freely in the planetary organization. Dynamic management principles and methods were introduced and used to speed up the changes needed to transform the company to a profitable enterprise.

At the start of the project in February 2000, the company had no commercial value. Four months after the project start the value of the company was calculated to be 32-44 MSEK or ca 3, 5-5 MUSD.

This is the first time this project has been made public, as the mother company was a public company on the Swedish stock exchange and the author was required to sign a long-term secrecy agreement.

1. Introduction

Other organizational models than the traditional ones have recently been of academic and practical interest, and management principles have been discussed and tested with a shift in focus from managerial behavior to management technology (Georges and Romme 2016). This includes so-called circular management practices, also known, for example, as holacracy (Robertson 2015a, b) and sociocracy (Buck and Villines 2007). Another model, originally developed for product development projects, maintaining management control but using dynamic principles and methods, is Planetary Organization (Stig Ottosson 1998b). By using planetary organization and dynamic management principles, interesting results have been achieved. The development of new products has, for example, shown that it is possible to do in a fraction of the time, compared to the use of traditional methods of organization and work (Stig Ottosson 2004, 2014). This paper discusses the experiences using Planetary Organization and dynamic principles to turnaround a company with 125 employees to profitability.

The development was done as an Insider Action Research (Stig Ottosson, 2003) project carried out by the first author as interim CEO and the second author as comet.

2. Theory

Excluded here as the information is more clearly treated in the book.

3. Management of Organizations

Excluded as the information is more clearly treated in the book.

4. The Turnaround Project

4.1 The project task

The turnaround project described here was to bring Frontec Research & Technology AB (FTR) in Gothenburg, Sweden to a profitable position within six months at an undefined cost payed by the business' own income and with support from the mother company, Frontec AB. Discussions between the chairman of the board and the first author about such a project started in December 1999. Prior to that, the mother company had tried traditional ways to change the negative curves without success. The discussions ended with the first author being given the mission, as interim CEO, to turn around the company. If he should not succeed, it was determined that the company was to be closed down.

When the project started, FRT had 125 qualified employees. It made a loss of about 0,9 MSEK per month, and had done so for the last six months. When the negative trend had started about one year earlier, different experts were consulted to change the situation. In early December, the CEO had decided to leave the company so the mother company to try something radical to change the situation.

The first author got a free hand to do what he thought best, without having to ask the board of the company for acceptance. Monthly financial reports had to be delivered to the mother company. The first author started his change management work in January 2000, although the formal start as CEO was on the 15th of February 2000.

4.2 From a line to a planetary organization

The company was at the start of the change management project, organized as a common line organization. In the last months before the project started, the company had lost about 35 skilled technicians, so the company now had, in total, 125 employees of which five were PhD's, four were Tekn.Lic (halfway between MSc and PhD), and 75 were MSc. There was one secretary for the CEO, one sales manager, one market manager, one finance manager, one IT manager and one receptionist. The customers were mainly larger Swedish companies, such as Volvo, Ericsson, AstraZeneca and Mölnlycke. Other customers were Sahlgrenska University Hospital and Chalmers University of Technology.

The change management (or turn around) project started on the 15th of February 2000, when all employees were first introduced to the new, interim CEO by the chairman of the FRT board. The new CEO then gave information about how he thought the turnaround should be done. He also informed them about the principles of a planetary organization and dynamic management. The three comets were introduced, as well as their mission.

One of the first symbolic actions of the interim CEO was to transform the big CEO office to a conference room, and he then took a small room as his office, since as his way of working was to be Management by Walking Around (MBWA). That was also how the Comets were to function—moving around talking with the associates and using mainly storytelling (Fryer 2003), in order to get them to reflect on other ways to work and to find solutions when they had problems to overcome. The interim CEO met at least once a week with the comets and the management team —the department managers—to find out what problems needed to be solved, as well to talk about problems that had been solved and lessons that had been learned.

The people in the administration got partly new topics to work with than those they had been used to. A business administrator was hired to act as an "early warning system" to avoid unwanted surprises—for the CEO to be one step ahead of the development, meaning to be pro-active instead or re-active. One important activity for the administrator was to go through contracts every week to find out in advance if any of the consultants were to be out of job in a near future. This was crucial because, on average, only 65% were doing paid work at the project start and that the ambition was to increase this to 80% within some months. The organization after the changes is seen in Figure 7.

Administration:
CEO (1 male)
Accounting (1 female)
Information (1 female)
Reception (1 female)
Early warning (1 male)

Deparments:
Strength calculation (30 men)
Production engineering (24 males + 1 females)
Product development (28 males + 2 females)
Software development (24 males + 2 females)
Web development (1 males + 2 females)

Comets:
Internal technician (male)
External HR consultant (male)
External technician (female)

Figure 7: The organization of FRT one month after the start of the turnaround project

The managers earned provision on how much was invoiced monthly for their consultancy work. That was changed so that a provision was shared equally by all managers, based on the total sales per month for the firm and not only on their "own" departments. Initially, this was not well received by the managers, but eventually it resulted in cooperation over borders of the departments. If one department had little to do and another had too much to do, people could automatically be used in a flexible way between the departments. In this way, the managers also got to know better what the colleges did and could take over their work, for example, in the event of sickness or vacation. Thus the managers started to help each other in a self-organizing way, resulting in more united management in the company and an increase of the collective knowledge in the firm.

4.3 Daily work

The daily work was based on the idea of making improvements every day. If a decision, in practice, showed to be less good or wrong, it was immediately changed. This can be seen as combining Lean principles and Pareto principles (Koch, 1998).

Everybody had to be hands on, so that even, for example, the CEO could pick up a paper in a corridor that should be in the litter box. However, routine work that took time from the work of the consultants, such as filling in travel expense forms, was transferred to the receptionist. She also made sure that the customers got monthly feed-back about the time the individual consultants had worked on different client projects. However, to get that feed-back from the consultants proved to be difficult. Therefore, the principle that they got reduced salaries until they had got into the habit of reporting their work in a proper way had to be implemented.

The managers were used to sitting in their offices a lot, doing less productive administrative tasks. Now they were encouraged to practice MBWA, meaning visits with their consultants, who generally worked in the clients' premises. They were also encouraged to ask the consultants to consider whether they could identify possibilities for improvement where they were working, giving the managers openings to discuss pilot improvement projects. This worked out well, and small initial jobs often developed into larger jobs involving more consultants.

The marketing and sales managers, who used to work with top people in the customer companies, had not been very successful in finding jobs for the consultants. They did not want to change strategy to work on middle-managers, who are closer to the daily work and can decide on new work to be done to speed up the outcome. Therefore, they were transformed into ordinary consultants, maintaining their managerial salaries for six to twelve months, which was possible based on how the employment contracts were written and signed. After some months, they left the firm, resulting in lower management costs as their positions were not occupied by someone else. A traditional lay off would have cost months of salaries during the termination time.

The financial reports that had to be delivered to the mother company every month did not show the core business costs. For example, management fees to the mother company were automatically added to the result. Therefore, a separate report for the CEO was introduced where such items were deducted. Based on these reports, the CEO now got clear information of how the core business of FRT was developing, making it possible to make fast adjustments when feasible.

4.4 The comets' input

The comets, who worked as mentors and used storytelling as a tool, encouraged, especially the younger consultants to speed up their work.

One example was a consultant making strength calculations on a special Table. He made a complete FEM (Finite Element Method) analyses. One of the comets came to his desk and asked him what he was working on, and then asked why he did not make simple calculations on most of the Table, concentrating the fine tuning on critical points. The consultant thought it over quickly and decided to make simple calculations on most of the Table and detailed analyzes on critical parts. This resulted in faster and cheaper results for a pleased customer, and also in a pleased consultant.

Another example was a consultant who did not want to give incomplete or imperfect Web solutions to the customer. If the client was not happy, she was unhappy and saw it as a criticism of her work and her professionalism. The comet told her about the Pareto principle, which meant that she could present solutions to the customer that were not yet final. This, in turn, resulted in engagement with and positive feed-back from the client.

4.5 Unexpected disturbances

Because of the problems in the company, the number of employees had the last two months before the change project started, and the personnel decreased from 160 to 125. As explained, different management consultants had been used the last year to try to find a formula to transform the company to be profitable again. Therefore, the employees were rather disillusioned and tired. In talking and with their body language, they demonstrated that they had lost faith and that they had very little hope for the future of the company. One example of that was that many employees did not hide that they spend much working time looking for new jobs. It was not uncommon that they left their computer screens on with job application forms on them.

Thus, initially, the great majority of the employees were indifferent to the changes, waiting to see the outcome doing their job with less enthusiasm and often with low efficiency. A few were openly negative and were asked to leave if they did not want to give the change project a try in a positive way. This was because their negative attitude would hamper the project.

Of the five managers, three were eager to tell the project leader that they were very positive to the way we were working and the principles we used. However, we soon got signs that they were not loyal and obstructed the work, for example, by spreading negative rumors that the mother company was in doubt of the new way of working and that money was flowing out of the company in an uncontrolled way, and both statements were false. As the rumors had been given face-to face, in confidence, to different employees, it was impossible to get evidences from them about the non-loyal behavior and why actions were difficult to perform to stop the dis-information and to get rid of the three managers. Later it turned out that the three managers had wanted to take over the company and that they even had made contact on this matter with the CEO of the mother company.

4.6 Result

After four months, the new way of organizing and working resulted in the possibility to invoice 85 % of the consultants' time. The company started to make a profit after two months of change, as is shown in Figure 8.

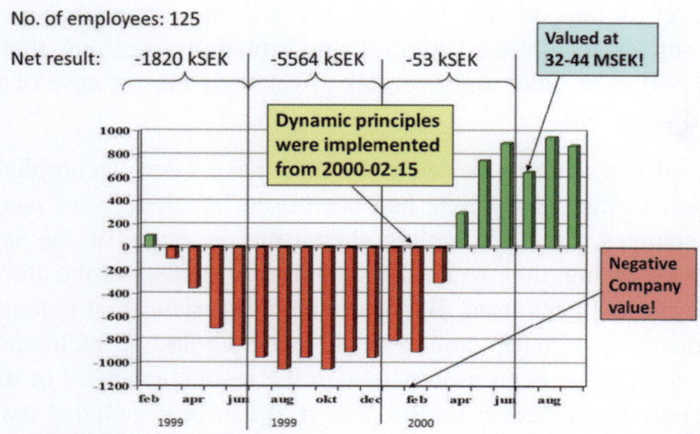

Figure 8: Statistics from before and after the management-change project of Frontec Research & Technology AB

In June 2000, the value of FRT was by independent evaluators estimated to be in the range 32-44 MSEK (Markkula and Engstrom 2000).

The registered official result at the tax authorities in Sweden for 1999 was −7 kSEK. For the period 1 January–30 June 2000 it was +29 kSEK.

5. Conclusions

The combination with the Planetary Organization concept and the use of dynamic methods and principles showed to be powerful in quickly getting the company showing blue Figures from its situation with heavy red financial Figures. The activities specified below were to be especially important to improve the result in this case:

- a “stripped-down” monthly financial report replacing the formal one for the mother company gave the CEO better possibilities to see how the core business was developing,
- an «Early Warning System» was introduced to take away unwanted surprises,
- the individual bonus system was changed for the managers to be a collective bonus system,
- continuous improvements (lean) and the Pareto principle were used intensively,
- MBWA was critical to pick up on weak signals and to improve the mood of the employees. MBWA also made it possible to meet the not confirmed bad rumors in a sophisticated way without revealing the knowledge of the rumors,
- the comets increased the efficiency of the work, based on storytelling and mentorship,
- by encouraging the consultants who were working at the premises of the customers to suggest small improvement projects and get feedback from the customers, the sales increased,

- the risk level for the company was lowered as each of the managers got to know the other managers duties, activities and situations meaning that they could better support each other and even take over their jobs in case of planned and unplanned situations.

Very few of the actions carried out could have been accomplished without difficulty if self-management alone had been used. As there were people working against the changes, officially saying they were in favor of the new ways of organizing and working, they would have hampered success of the activities if pure self-management had been used. Based on these experiences, it is hard to see how self-management, a circular and/or a holacracy/holacentric/circular/sociocratic organization could have been used to obtain the results achieved in this case.

Based on our experiences from the project, it can be concluded that neither the line or matrix organizations nor self-management organizations as circular, holacracy, holacentric, or sociocratic are suited for handling modern, complex societal demands and fast moving changes. In this case, the planetary organization worked. To find out if the findings from this case can be generalized, more projects need to be done and followed. This seems to be especially important as traditional management methods and techniques are not designed to handle instability and rapidly changing situations.

About the Authors

Stig Ottosson is a PhD in Physics (Chalmers University of Technology in Sweden), an MBA (Gothenburg University in Sweden) and a DSc (Docent*) in Innovation Technology (Linköping University in Sweden). He has more than twenty years of experience in managing new product and innovation development in different branches and sectors. For example, he was responsible for SKF New Products, which was a significant business success for the multinational bearing company SKF. In 2008, he founded Öckeröarna Rotary Club in Sweden and was the first (charted) president. In 2010, he also founded Göteborg–Lindholmen Rotary Club. Stig has also fourteen years' experience in Sweden, Germany, and Norway as a Professor in Product Development (Halmstad University and Otto-Von-Guericke University) and Innovation Management (Linköping University). Since 2012, he is professor in Innovation Management at Norwegian University of Science and Technology (www.ntnu.edu). He is on the editorial boards of number of scientific journals and the scientific boards of various international conferences such as the Design Conference in Dubrovnik, the International Conference if Engineering Design (ICED), Design Conference, and the Tools and Methods in Competitive Engineering Conference (TMCE)—being held biennially at various locations around the world.

Evastina Björk is Master in Occupational Therapy (Jönköping University in Sweden), a PhD in Product Development (Otto-von-Guericke University in Germany) and a DSc (Docent*) in Public Health Science in the topic universal design at the Nordic School of Public Health in Gothenburg, Sweden. Since 2012, she is associate professor in Universal Design at Norwegian University of Science and Technology. Her research area of interest is universal design and welfare technology, meaning investigating how cross professional cooperation in organizations have effects on the outcome of universally designed welfare-technology solutions. She is also involved in different international conferences such as the Design Conference in Dubrovnik, the International Conference if Engineering Design (ICED), Universal Design Conference and is biennially at various locations around the world. She has experience of project management from several industrial and academic projects over the years.

Docent: The Swedish degree of Docent requires, in addition to a PhD degree or equivalent, documented abilities in scientific research beyond PhD work and documented pedagogical competence at the advanced graduate level. The Docent is obligated to serve as opponent/ examiner at the public defense of a doctoral thesis.*

References

Barnevik, P. (2011). *Jag vill förändra världen*: Albert Bonniers Förlag.

Björk, E. (2003). *A Contribution to Insider Action Research: Applied on Development of Assistive Products*: E. Björk.

Buck, J., & Villines, S. (2007). *We the people: Consenting to a deeper democracy: A guide to sociocratic principles and methods*: Sociocracy. info.

BusinessDictionary. (Ed.) (2010).

Crumley, C. L. (1995). Heterarchy and the analysis of complex societies. *Archeological Papers of the American Anthropological Association, 6*(1), 1-5.

de Vries, M. F. K., & de Fries, M. F. K. (1989). *Prisoners of leadership* (Vol. 36): Wiley New York.

Döner, A. S. (2016). Organizational Forms: Knowledge, motivational and power dynamics.

Fryer, B. (2003). Storytelling in Business. Retrieved from

Georges, A., & Romme, L. (2016). *Management as a science-based profession: A grand societal challenge*. Retrieved from Eindhoven

Gladwell, M. (2000). *The Tipping Point*. London: Little, Brown and Company.

Handelshögskolan. (2014). Handelsalumnen Percy Barnevik mötte dagens studenter. *Handelshögskolan*. Retrieved from http://handels.gu.se/om_handelshogskolan/Nyheter/fulltext/handelsalumnen-percy-barnevik-motte-dagens-studenter.cid12

Holmström, B. (1999). Managerial incentive problems: A dynamic perspective. *The Review of Economic Studies, 66*(1), 169-182.

Imparato, N., & Harari, O. (1996). *Jumping the curve: innovation and strategic choice in an age of transition*. San Francisco: Jossey-Bass.

Koch, R. (1998). The 80/20 principle. The secret of achieving more with less.

Krohn, W., Küppers, G., & Novotny, H. (1990). *Portrait of a scientific revolution*: Springer.

Markkula, & Engstrom. (2000). Företagsvärdering av Frontec Research & Technology AB. Retrieved from

Marshall, J. C., & McHardy, B. (1999). *The Principles of Self-management: the Key to Personal and Professional Success*: Selecting Sales Professional.

Ottosson, S. (1996). Dynamic product development: findings from participating action research in a fast new product development process. *Journal of Engeering Design, 7*(2), 151-169.

Ottosson, S. (1998a). Planetary Organisation Offers Advantages in Project Work *Designers* (pp. 196-201): Springer.

Ottosson, S. (1998b). Planetary organizations. *Technovation, 19*(2), 81-86.

Ottosson, S. (2003). Participation action research-: A key to improved knowledge of management. *Technovation, 23*(2), 87-94.

Ottosson, S. (2004). *Stage-Gate™ vs. DPD in software development.* Paper presented at the IPD 2004: Proceedings of the 5th Workshop on Integrated Product Development, Magdeburg, Germany, 22.-24.09. 2004.

Ottosson, S. (2009). *Frontline Innovation Management.* Sweden: Tervix.

Ottosson, S. (2016). *Developing Sustainable Product Innovations.* Sweden: Tervix.

Ottosson, S., & Björk, E. (2000). Participation action research—a method to improve knowledge of product development and innovation management. *Tools and Methods in Competitive Engineering (TMCE-2000).*

Robertson, B. J. (2015a). *Holacracy: The new management system for a rapidly changing world*: Macmillan.

Robertson, B. J. (2015b). *Holacracy: The revolutionary management system that abolishes hierarchy*: Penguin UK.

Romme, A. G. L., & Endenburg, G. (2006). Construction principles and design rules in the case of circular design. *Organization science, 17*(2), 287-297.

Romme, G. (2016). *The Quest for Professionalism: The Case of Management and Entrepreneurship*: Oxford University Press.

Salton, G. J. (2014). 3-Dimentional Organizational Charts. Retrieved from http://garysalton.blogspot.no/2014/09/3-dimensional-organizational-charts.html?m=1

Wall, C., & Westmar, J. (2007). *Kamprad vs. Wallander Tolkningsparadigm på svenskt entreprenörskap.* (Master), Lund University, Lund.

Appendix C
3D-Printing as a Creative Modelling and Prototyping Tool

Paper accepted for the AHFE 2018 conference, Loews Sapphire Falls Resort, Universal Studios Orlando™, Florida, USA, July 21–25, 2018

Stig Ottosson, Professor (Norwegian University of Science and Technology), Norway

Abstract
3D-printing offers possibilities to quickly and cheaply play with different body shapes, material texture, and the functions of objects as well as to test, compare and judge which of the different iterations to settle on or to combine as a final solution. In the development of new tangible (hardware) products, creative modelling can also be used to obtain retailer, user and customer opinions of new product concepts. To experience using 3D-printing as a modelling tool, we made a strict copy of simple plate product from which we could "play" with different creative solutions.

We found that realizing extended product functionality that gradually emerged in the creative process had been (extremely) expensive and long-lasting using classical modelling and prototyping methods because complicated tools had had to be manufactured and changed before each test iteration. Using 3D-printing, it was easy to make changes, and new ideas could easily be tested, which increased not only the functionality but also other product values.

Introduction
This paper shares some experiences gained from a new product development (NPD) work carried out, for which 3D-printing was used as a creative modelling and prototyping tool. The mission was to exchange a simple metal plate cover of a safety belt buckle with a plastic cover that had more functionality and that possibly could be produced with additive manufacturing or with traditional plastic-forming production methods.

S. Ottosson, *Developing and Managing Innovation in a Fast Changing and Complex World*, https://doi.org/10.1007/978-3-319-94045-8

Before the work started, the problem was to find a simple product to improve. However, the author had contacts with the Swedish company Careva AB that is acting in the welfare technology sector. One product line of the company is positioning belts. These belts keep disabled passengers in vehicles and wheel chairs in an upright sitting position (see Figure 1). Without the positioning belts, the function of the safety belts in vehicles is poor and dangerous traffic situations can occur. In addition, a problem with the positioning belts is that children can open the locks when being transported. Therefore, the company sells a black-painted metal cover that is pushed over the locks preventing such unwanted unlocking.

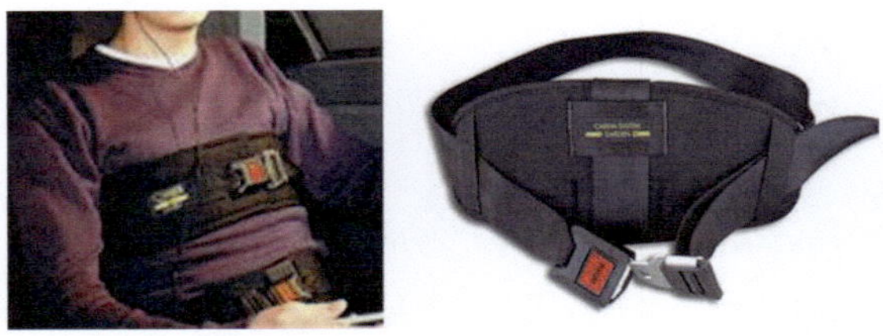

Figure 1: An example of positioning belts for disabled passengers in vehicles and wheelchairs (from www.careva.se)

When the metal cover is pushed over the belt locking, the lock can be opened by pushing, for example, a ballpoint pen through the ø5 mm hole that is situated over the pushdown release button (see Figure 2). The hole is small enough to prevent even small fingers from releasing the release button. The metal cover is heavy and is a rather rough solution, lacking, for example, a klick sound when the cover is in the right position over the lock. Therefore, the author decided to study the process of developing a new, "better" cover using 3D-printing as a mediating tool.

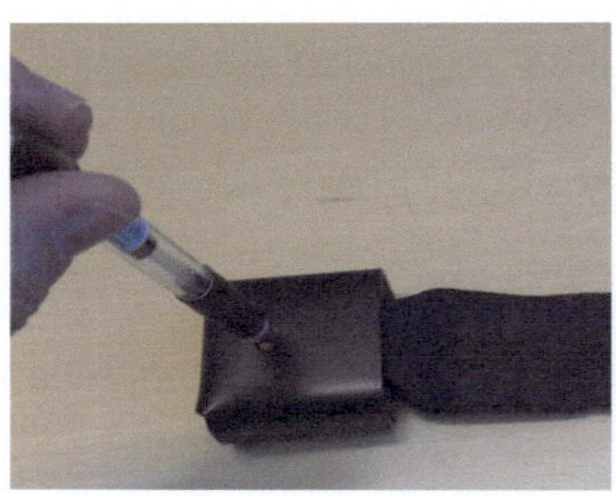

Figure 2: Through the hole in the cover is an opening device – here a ballpoint pen is used to unlock the belt

The initial part of the work was earlier reported as a conference report (Ottosson and Skogsrød 2016).

Research Method used

The author acted as project leader and used common Action Research principles (Ottosson 2003) for the scientific part of the work.

Definitions

To find and produce new solutions, *creativity* and *improvisation* are needed by the people doing the development work. These two terms can be defined as follows.

> "*Creativity* means shakings things up, both inside ourselves and in the world around us, and the constant re-organizing of both cognitive schemata and, to a greater or lesser extent, the domain of the creative person's activity." (Montuori 2003)

> "*Improvisation* is thought of as making the best of things, while awaiting a return to the way things should be done. Improvisation is an exception, something we can 'fall back on' when things do not go the way they should." (Montuori 2003)

In the development of new tangible (hardware) products, creative modelling can be used in the chain from the *ideation* (idea generation) to the testing of user and customer opinions of new product concepts. Practically seen, *creative modelling* means iteratively making and testing models and prototypes so that the end result can be new sustainable innovations that meet a large number of product values.

There are many definitions of the terms "models" and "prototypes". In this paper, a *model* will be a formed, solid object that demonstrates the function and/or the shape of the product. The model can also be a part of a product concept used in the further development of a new product. When a model exists, a prototype can be developed, which means the development of a production-ready solution is developed. The definition of a *prototype* is "an approximation of the product along one or more dimensions of interest (Ulrich and Eppinger 2016)". In this study, the development work to make a model is called modelling and the development of a prototype is called prototyping.

Models are important visualization tools for the developers, as well as for managers, users, and customers. When one is able to touch a model, one gets a firmer mental picture to work from than if only a picture or digital mock-ups exist. Thus, models help to provide a better understanding of the priorities and help us to avoid mistakes and misunderstandings.

It is sometimes recommended that the modelling should be simple and rough. The strength of rough modelling with simple materials is that it encourages playing with ideas (Ottosson 2009). Simple materials, such as clay, paper, or wood are quick and easy to change, which is why they should be used preferably before using harder materials. "When a model starts to harden up, so also does the thinking." (Schrage 2000, p. 79). Note that modelling with "mock-ups", used in the shaping of new products, do not generally demonstrate the function of the products.

Another effect of modelling and prototyping is that both tacit and explicit knowledge is used on an individual and on a collective basis. “Using prototypes and prototyping may hold a monumental potential to better capture and transfer knowledge in product development, thus leveraging existing integration events in engineering as a basis for knowledge transformation” (Erichsen et al. 2016).

Although the development work can be based on continuous improvements of an existing product, which is called Lean product development, sometimes—if the developers are creative and can improvise—radical new solutions can occur, providing a possible foundation for radical innovations.

Development theory

Left out as it is treated in the book.

From idea to model and prototype

The start of the development work was to make a CAD design that closely related to the metal cover. Then we modelled it to be suited for 3D-pringing in our university machine, which is an EOS P395 machine. (The powder used was PA2200 and the belt buckle was driven by a standard EOS balance parameter.)

Therefore, the first 3D-printed model in our NPD project was a strict copy of the functional metal cover (see Picture 2 in Figure 4). The first creative idea that came to mind was that the product name, protruding 0.5 mm, could be done so that it would add to the image value of the product and company (see Picture 3 with a painted upper surface in Figure 4). Also, the shape of it could be done to closely resemble the lock itself, which would increase the sensorial and image values (see Picture 4 in Figure 4). Picture 5 in Figure 4 is a small change in the shape of the cover to get it look better. With that change, the original design in Picture 1 had been left behind. The next step was to move the hole (Picture 6 in Figure 4) as we realized that by doing so, the release power could be lowered, providing a functional improvement. To lessen the environmental impact and make it a sustainable solution, we shortened the length of the cover (see Picture 7 in Figure 4).

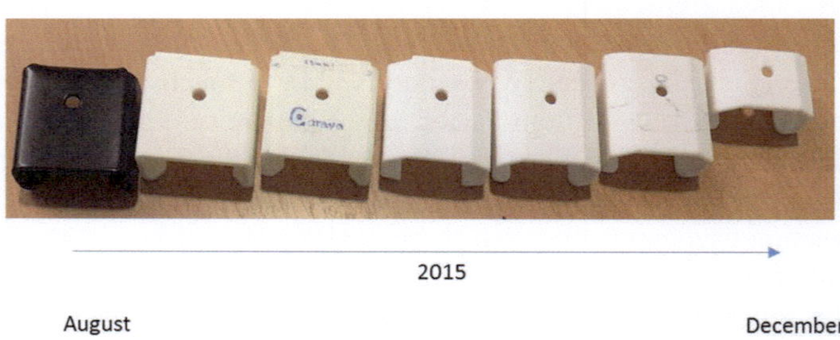

Figure 4: The development chain from the original metal lock to a functional prototype

Then, as a creative idea, we added a thin line protruding 0.2 mm under the lock to ensure that the cover did not unintentionally slide off the locking (see Figure 5). That solution gave an unintended clicking sound indicating that the cover fitted correctly when pushed over the lock. In product value terms, this meant an additional functional as well as sensorial value to the product.

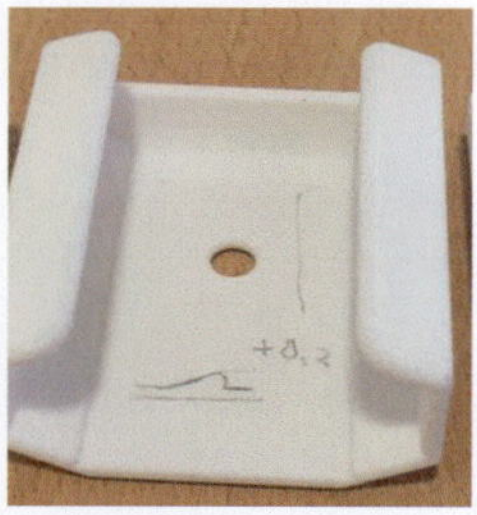

Figure 5: The first idea to add a clicking and holding line under the cover

To change the holding force and to get a higher clicking sound, our tests showed that the line had to protrude 0.5 mm to achieve the desired requirements on the clicking sound and ensuring that the cover did not fall off when the lock was shaken.

One negative effect of using the cover is that children may feel imprisoned or claustrophobic, which can be a problem for the existential design to overcome (Torskildsby 2014). However, in January 2016, when we were planning a lecture in product development, the design professor Kari Øverseth at NTNU in Gjøvik got an impulse to make a quick sketch showing that the cover could be made into a "cat shape" to please children (see left Figure 6). Therefore, we sketched the idea on the cover (see right Figure 6) for the CAD work to be done before we could make next model.

Figure 6: The left sketch was transformed to instructions on how to realize the idea in reality

The addition of the cat on the cover has no functional value, but it can be used in a positive way to take away claustrophobic feelings for small children. For example, parents can say that they will ask the cat if it is OK to lock or release the child. It is worthy of note that the short, functional cover in Figure 6 did not give a pleasant aesthetic look, although the shorter cover meant a more sustainable solution as 35% less material was used. We therefore went back to the original length of the cover. Figure 7 shows the result and how we thought it feasible to let the parents paint the cat with a color they preferred in a simple way.

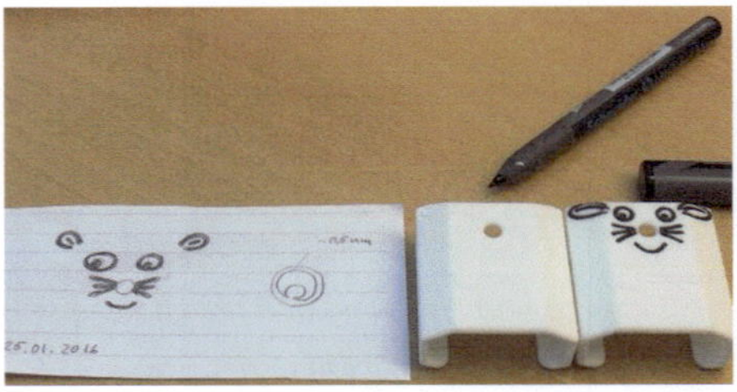

Figure 7: The end result of the development experiment showing the PAD-draft, the 3D printing outcome and the result after a marker pen had been used to fill in the "ditches"

However, the company (www.careva.se) did not want to prioritize small children with the product shown in Figure 7. The reason was that the most primary users for them are youngsters and adults, for which the product should be as "invisible" as possible. This is why they now expressed that they wanted the product to be black with a soft surface, much like the powder-painted black metal cover in Picture 1 in Figure 4. If possible, they wanted the product to be shown at the Reha Care Fair in Düsseldorf, Germany, in August 2016.

As there was no known firm in Norway that could make black smooth 3D-printings, we asked the Swereco experts in Göteborg (www.swereco.se) for suggestions of serious companies in the field and received a list of contacts in Sweden and Poland. A Swedish company gave us the best impression from their homepage and contacts over telephone. They claimed they could not make black covers, but suggested SLS (Selective Laser Sintering) details in polyamide. In an email from them, they suggested black dyed products, which "are basically white, but dyed black. The staining penetrates a small piece of the material and is more durable than lacquer." They did not want to make a test piece free of charge, but Careva ordered ten pieces on 10 October.

After they had received the order, they sent a message by email stating that "your file must be adjusted so the "Careva" logo will be at least 0.6 mm thick." That sounded a bit strange, as we had had no problem with that in the printing with our university machine. However, we changed the CAD file and they printed, dyed the products black, and delivered them on 27 October 2016 (see Figure 8). However, the result was disappointing with regard to the surface structure. (The machine they used was an EOS P760 and the material used was PA2200.)

Figure 8: The product with a painted C. The surface is like an orange and frosted, which is difficult to see from the photos

When we complained about the color and surface finish, we got an email message on 28 October: "If it is a smooth finish you are after, we would have to putty and polish more, and then I do not think this method would be profitable. Vacuum casting would be a better option as you can get black material from the start." They also told us that they do this type of production.

We had come to a dead end and did not know how to proceed if we did not accept vacuum forming of the products, which would mean that we had capitulated, so the project was put on hold.

On the 8 November 2016 the author attended the weekly meeting of the Öckeröarna Rotary Club, at which a representative of Magicfirm Europe told about their 3D-printing development. He sent around some difficult-to-produce samples of a toy train engine and told us that the company was working on mixing different colors. The smoothness of the surface was considerably better than what we had gotten in Norway and from the Swedish firm. He promised to make a sample free of charge, which arrived on 30 November 2016 (Figure 9).

Figure 9: The first sample from Magicfirm Europe, printed in their machine ZYYX+ 3D Printer

The hole on the sample was not perfect and the surface inside it was not as good as on the outside. We queried this and received their answer on 9 December 2016:

"we believe we can improve the surface quality, but we think we would need to make some slight modifications to the model in order to do so. Our machines are capable of printing in multiple colors, and more information on this process can be found at: www.zyyx3Dprinter.com/print-unlimited-colors-and-unlimited-materials-with-your-zyyx-3D-printer/". To do that, they asked for quite a large sum of money without giving any guarantees that the surface demands could be met.

Based on the negative experiences from our struggling to get an Additive Manufacturing solution, Careva decided it was too risky to continue the project and thought the investment in plastic tooling had to be pushed on the future, although the prototype in Picture 7 in Figure 4 and in Figure 5 showed pleasant product values except from the surface smoothness.

Discussion

The use of 3D-printing made it possible to satisfy all the product values mentioned in the theory part of the paper, the ISO (1998) demands, the usability demands, and the 7+2 UD-principles. Thus, all the product demands in the Definition section were satisfied in the development of this product.

The development of the new product model was based on an existing solution, which is why BAD-PAD was only used in a limited way. Figure 10 shows the sequences used. That a functional solution existed from the start contributed to the appearance of the so called anchoring (or focalism) effect and delayed the development in the form and indicated that more steps needed to be taken.

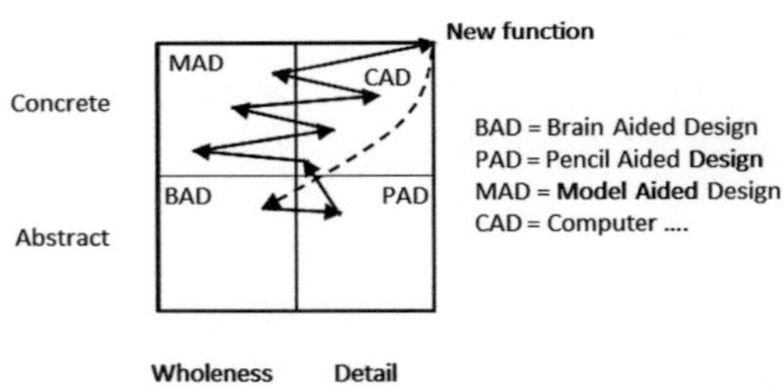

Figure 10: How the design activities were done in the test example

In this case, the creativity was felt to be higher than what could possibly have been the case if 3D-printing had not been used. The main reason for this conclusion is that there were few—if any—limitations as how the product could be designed in terms of interior and exterior shape. Another reason was the easiness in testing new ideas that evolved when each new model had been produced.

The functionality that gradually emerged in the creative process had been (extremely) expensive and long-lasting to realize with classical prototyping methods as in that case, complicated tools had had to be manufactured and changed before each testing iteration.

Using our method, it was easy to make changes, and new ideas could easily be tested, which increased not only the functional but also the other product values.

The surface roughness is at a level that makes Additive Manufacturing not useful for the production of this product. Going to other production machines would require new tests and the functional values might change. Another limitation is that it is not easy to get different colors other than white or grey. Painting did not give an acceptable result in terms of wear, and coloring using chemical bath procedures is questionable from a sustainability point of view. However, if white or grey colors can be accepted, this product demands less material than when it is produced in other ways, which influences the environmental aspect from the cradle to the grave of the product.

In our case, it turned out to be more problematic to go from the 3D-printed prototype to Additive Manufacturing than anticipated because the production machine could not make the thin lines made by the prototyping machine. Redesign had to be done to meet that demand. The surface roughness meant that the product did not meet the specifications.

One effect of the problems with the transfer from the prototype machine to the production machine was that time to market (TTM) was delayed. Frustration could have been avoided and time could have been gained if we had found out about the surface qualities and minimum line-widths of different production machines in the development stage of the prototype. We had received plate samples with good surface quality and in different colors. We regarded these as samples from the machine, when, in reality, they must have been mass-produced samples, most probably from a plastic granule provider. This can be seen as a miscommunication between us and the provider. A visit with the producer could have revealed this.

Conclusions

From this project, we have learned that modelling and prototyping with 3D-printing is beneficial when many functions are to be fulfilled in NPD. Thus, as a creative modelling and prototyping tool, 3D-printing showed in this case to give useful results and also to give unforeseen positive effects such as the clicking sound.

We have also learned that going from prototype to production was much more difficult than anticipated, causing in the end a dead end in the project. Because false information from the producer, time to market (TTM) was severely delayed and eventually infinitely long and as the firm lost confidence, and market feedback could not be gained. The earnings disappeared when the whole market introduction was eventually stopped.

It is clear that for similar cases, it is important to find out early about the machine data of different production machines and to visit the company or companies to verify that they can, in fact, deliver the required products.

The 3D-printed covers could have been used to get dealer, user and customer feed-back. However, for an eventual production, the surface roughness is not good enough, which is why other plastic manufacturing methods have to be considered.

The DPD™ method was shown to be useful, but the BAD-PAD-MAD-CAD model needs an adjustment when modelling is based on an existing solution.

References

Erichsen, J. A. B., Pedersen, A. L.; Steinert, M. & Welo, T. (2016): Using Prototypes to Leverage Knowledge in Product Development: Examples from the Automotive Industry. I: IEEE International Systems Conference (SysCon 2016) Proceedings, Orlando, Florida: IEEE 2016 ISBN 978-1-4673-9518-2. s. 491–496

ISO 9241-11:1998(en): https://www.iso.org/obp/ui/#iso:std:iso:9241:-11:ed-1:v1:en

Montuori, A. (2003): The complexity of improvisation ant the improvisation of complexity: Social science art and creativity. Human Relations Vol 56(2), p 237

Ottosson, S. & Skogsrød, T. (2016): Some Experiences from Using 3D-printing in the Modelling of a New Product, NordDesign 2016, Trondheim, Norway, August 10–12

Ottosson, S. (2018): Managing and Developing Innovation in a Fast Changing and Complex World, Book at review at Springer Science

Ottosson, S. (2009): Frontline Innovation Management, Tervix AB, ISBN 978-91-977947-7-0

Ottosson, S. (2003): Participation Action Research—A Key to Improved Knowledge of Management, Technovation - the International Journal of Technological Innovation and Entrepreneurship, Vol 23, pages 87–94

Schrage, M. (2000): Serious Play - How the World's Best Companies Simulate To Innovate, Harvard Business School Press

Story, M.F., Mueller, J.L., Mace, R.L. (2001): The Universal Design File: Designing for People of all Ages and Abilities, Raleigh, North Carolina: North Carolina State University

Torkildsby, A B. (2014): Existential Design – Revisiting the "dark side" of design thinking, PhD thesis at University of Borås, Sweden

Ulrich, S.D. & Eppinger, K.T (2016) (6th edition): Product design and development, McGraw-Hill Education, NY

Vajna, S. and Kittel, K. (2009): An Approach to Compare Product Development Methods, In DS 58-5: Proceedings of ICED'09, the International Conference on Engineering Design, Vol. 5, Design Methods and Tools (pt. 1), Palo Alto, CA, USA, Appendix 3: 3D-printing as a Creative Modelling and Prototyping Tool

References

Adams, R. S., Turns, J., & Atman, C. J. (2003). Educating effective engineering designers: The role of reflective practice. *Design Studies, 24*(3).

Agha, R. A., Fowler, A. J., & Sevdalis, N. (2015). The role of non-technical skills in surgery. *Annals of Medicine and Surgery, 4,* 422–427.

Altshuller, G. S. (1980). *Creativity is an exact science*. New York: Gordon and Breach Science Publishers.

Aurisicchio, M., Bracewell, R., & Wallace, K. (2006). Characterising the information requests of engineering designers. In *International Design 2006 Conference, Dubrovnic*.

Alströmer, C. (1998). Private communication about the writing of manuals at Ericsson Corp, where Claes Alströmer had been project manager.

Andreasen, M. M., & Hein, L. (1987). *Integrated product development*. London: Springer.

APM. (2000). *Project Management Body of Knowledge,* Association for Project Management, M. Dixon.

Baker, M. (2008). *The strategic marketing plan audit* (p. 3). ISBN 1-902433-99-8.

Balmer, J. M. T. (1997). *Corporate identity: Past, present and future*. International Centre for Corporate Identity Studies, Working paper series 1997/4.

Barnevik, P. (2011): *Jag vill förändra världen.* Albert Bonniers Förlag. ISBN: 9789100126612.

Björk, E. (2003). *Insider action research applied on development of assistive products*. PhD thesis, Otto-von-Guericke-University, Magdeburg, Germany.

Björk, E. (2009a). Lessons learned from a want based NPD project. In *International Design Conference—Design 2008, Dubrovnik.*

Björk, E. (2009b). Why did it take four times longer. *International Journal of Technology and Disability, 21*(4), 159–170.

Bjerke, B (2013). *About entrepreneurship.* Edward Elgar Publishing. ISBN 978 1 78254 5385.

Björk, E., & Ottosson. S. (2007). Aspects of consideration in product development research. *Journal of Engineering Design, 17*(4).

Björk, E., & Ottosson, S. (2008). Lessons learned from a want based NPD project. In *International Design Conference—Design 2008, Dubrovnik, Croatia*, May 19–22.

Björk, E. & Ottosson, S. (2016). Cross-professional Cooperation in a university setting. In *Universal design 2016: Learning from the past, designing for the future.* ISBN 978-1-61499-683-5 (print) | 978-1-61499-684-2 (online).

Björk, E., Ottosson, S. &Thorsteinsdottir, S. (2008). *E-learning for all.* In A. R. Lipshitz et al. (Eds.), *E-learing—21 century issues and challenges* (pp. 47–68). New York: Nova Science Publishers. ISBN 978-1-60456-156-2.

S. Ottosson, *Developing and Managing Innovation in a Fast Changing and Complex World*, https://doi.org/10.1007/978-3-319-94045-8

Branzell, A. (1995). *Något om rummet (in Swedish): Architecture*. Göteborg, Sweden: Chalmers University of Technology.

Burke, R. (2011). *Project management techniques*. Burke Publishing. ISBN-13:978-0-9582733.

Böhmer, A., Beckmann, A., & Lindemann, U. (2015). Open innovation ecosystem—Markerspaces within an agile innovation process. In *ISPIM Innovation Summit, Brisbane, Australia*, December 6–9.

Carr, N. G. (1999). Marketing: The economics of customer satisfaction. *Harvard Business Review*, 15–18.

Chakravarthy, P. (2011). *The difference between coaching and mentoring*. Forbes 20th December. http://www.forbes.com/sites/infosys/2011/12/20/business-leadership-for-smarter-org-2/#619e9553406f.

Chen, I. J., & Paulraj, A. (2004). Towards a theory of supply chain management: The constructs and measurements. *Journal of Operational Management, 12,* 119–150.

Cialdini, R. B. (2000). *Influence: Science and practice* (4th ed.). Ohio: Allyn & Bacon. ISBN 0-688-12816-5.

Clifford, C. (2014). *Entrepreneur*. Entrepreneur Media, Inc. Retrieved May 25, 2014.

Cooper, R. G. (1994). Perspective third generation new product processes. *Journal of Marketing Management, 11,* 14–43.

Cooper, R. G. (2001). *Winning at new products—Accelerating the process from idea to launch*. New York: Basic Books.

Cunha, M. P., & da Cunha, J. V. (2002). To plan and not to plan: Toward a synthesis between planning and learning. *International Journal of Organization Theory and Behavior, 5*(¾), 299–316.

Cusumano, M. A. (1997). How Microsoft makes large teams work like small teams. *Sloan Management Review, 1997,* 9–20.

Cusumano, M., & Selby, R. (1995). *Microsoft secrets*. New York: The Free Press.

Dreyfus, H. L., & Dreyfus, S. E. (1986). *Mind over machine: The power of human intuition and expertise in the era of the computer*. Oxford: Basil Blackwell.

Döner, A. S. (2016). *Organizational forms: Knowledge, motivational and power dynamics*. The Academic Elegance. www.researchgate.net.

Engholm Jensen, K., Jensen, J. P., Digmann, A., & Bendix, H. W. (2008). *Principper for offentlig innovation - fra best practice til next practice*. Copenhagen: Börsens Forlag. (in Danish). ISBN 978-87-7664-196-2.

Engwall, M. (1995). *Jakten på det effektiva projektet*. Stockolm, Sweden: Nerenius & Santérus Förlag. (in Swedish).

Erikson, T. (2017). *Omgiven av idioter*. Månpocket. ISBN 978-91-7503-725-7 (in Swedish).

Farner, A. (2008). *Verksted verktöj – Å planlegge of lede som workshops*. Oslo: Kommuneforlaget. (in Norweigan).

Flin, R., O'Connor, P., & Crichton, M. (2012). *Safety at the sharp end*. Tokyo: Kaibundo Publishing Co.

Garvey, B., Stokes, P., & Magginson, D. (2014). *Coaching and mentoring: Theory and practice*. London: SAGE Publications.

Geoghegan-Quinn, M. (2013). *Citation from a speech*. Sweden: Dagens Industri. February 07, 2013.

George, J. M. (1996). Trait or state affect. In K. R. Murphy (Ed.), *Individual differences and behavior in organizations* (pp. 145–171). San Francisco: Jossey-Bass.

Gilmore, R. (1995). Alice in the quantum land. *Copernicus*. ISBN 978-0387914954.

Gillberg, M. (1999). *From green image to green practise*. Sweden: Lund University, Lund Studies of Law.

Gladwell, M. (2000). *The tipping point*. London: Abacus.

Gourville, J. T. (2006). Eager sellers and stony buyers. *Harvard Business Review*.

Gummesson, E. (2002a). *Total relationship marketing: Rethinking marketing management*. Butterworth-Heinemann Ltd. ISBN 0750654074.

Gummesson, E. (2002b). Relationship marketing and a new economy: it's time for de-programming. *Journal of Services Marketing, 16*(7), 585–589.

Gustavsson, B. (2001). *Kunskapsfilosofi, Tre kunskapsformer i historisk belysning (in Swedish).* Smedjebacken, Sweden: Fälth & Hassler.

Hackman, J. R. (2002). *Leading teams: Setting the stage for great performances*. Harvard Business school Press. ISBN 1-57851-333-2.

Hamm, S. (2007). *Bangalore tiger*. New Delhi: Tata MacGraw-Hill.

Handelshögskolan. (2014). *Handelsalumnen Percy Barnevik mötte dagens studenter.* Handelshögskolan. Retrieved from http://handels.gu.se/om_handelshogskolan/Nyheter/fulltext/handelsalumnen-percy-barnevik-mottedagens-studenter.cid12.

Haynes, B., & Price, I. (2004). Quantifying the complex adaptive workplace. *Facilities, 22*(1/2), 8.

Henkel, J., & von Hippel, E. (2005). Welfare implications of user innovation. *Journal of Technology Transfer, 30*(1/2), 73–87.

Hesselbein, F., Goldsmith, M., & Somerville, I. (Eds.). (2002). *Leading for innovation and organizing for results*. San Francisco, CA: Jossey-Bass.

Highsmith, J. (2002). *Agile software development ecosystems*. New York: Addison-Wesley.

Highsmith, J. (2004). *Agile project management: Creating innovative products*. New York: Addison Wesley. ISBN 0321219775.

Holm, M., & Zhang, L. (2016). *Swedish and Chinese Leadership at CEVT—A case study on the cultural differences of Chinese and Swedish leadership*. Report No. E2016:039 from Chalmers University of Technology, Department of Technology Management and Economics, Gothenburg, Sweden.

Holmdahl, L. (2007). *Complexity aspects of product development*. Ph.D. thesis, Otto-von-Guericke-University, Magdeburg, Germany.

Holmdahl, L. (2010). *Lean product development på Svenska*. Göteborg: Holmdahl Förlag. ISBN 978-91-979196-0-9.

Holtskog, H., Carayabbus, E.G., Kaloudis, A., & Ringen. G. (2018). Learning factories—the nordic model of manufacturing. In *Palgrave studies in democracy, innovation, and entrepreneurship for growth*. ISBN 978-3-319-41886-5.

Hoque, F. (2013). *Everything connect—How to transform and lead in the age of creativity*. Innovation and Sustainability: McGraw Hill.

Hubka, V. (1995). *DESIGN FOR – DF. Fertigungsgerechtes Konstruieren* (pp. 1–6). Paper presented at the Beiträge zum 6. Symposium, Erlangen Germany.

ICC/ESOMAR. (2008). *International code on market and social research*. ICC/ESOMAR Amsterdam, the Netherlands, 4th ed. http://www.esomar.org/uploads/pdf/professional-standards/ICCESOMAR_Code_English (2nd ed.). SAGE. ISBN 978-1-4129-1319-5.

Imperato, N., & Harari, O. (1996). *Jumping the curve: Innovation and strategic choice in an age of transition*. San Francisco: Jossey-Bass Publishers.

Industrial Psychology Research Centre. NOTSS. (2012). *Non-technical skills for surgeons* [online]. Available at: http://www.abdn.ac.uk/iprc/notss/. Accessed April 14, 2012.

ISO 9241-11:1998(en). https://www.iso.org/obp/ui/#iso:std:iso:9241:-11:ed-1:v1:en.

Jackson, S., Schwab, R., & Schuler, R. (1986). Toward an understanding of the burnout phenomenon. *Journal of Applied Psychology, 71*(4), 630–640.

Johannisson, B. (1996). Entreprenörskap – gränslöst ledarskap. In G. Ekvall (Ed.), *Navigatör och inspiratör*. Lund, Sweden: Studentlitteratur. (in Swedish).

Jordan, P. N. (1998). Human factors for pleasure in product use. *Applied Ergonomics, 29*(18), 25–33.

Johnstad, T., Giaever, T, Holtskog, H., & Strand, T. (2012). *LEAN på Norsk – med industrierfaringenr fra Rausfoss-industrien*. Vallet: Opplandske Bokforlag. ISBN 978-82-7518-201-0 (on Norwegian).

Järrehult, B. (2006). *Presentation at World Tissue Manufacturing Symposium, Karlstad, Sweden*.

Kennedy, M. (2003). *Product development for the lean enterprise—Why Toyota's system is four times more productive and how you can implement it*. Richmond, Virginia: The Oaklea Press.

Kennedy, M., Harmon, K., & Minnock, E. (2008). *Ready, set, dominate—Implement Toyota's set-based learning for developing products and nobody can catch you*. Richmond, Virginia: The Oaklea Press.

Kanter, R. M. (1983). *The change masters: Innovations for productivity in the American Corporation*. New York, USA: Simon & Schuster.

Karlsson, E. (2008). *Leveransalternativ för e-handel med dagligvaror*. PhD thesis from School of Economics, Göteborg University, Sweden (in Swedish).

Kodate, N., Ross, A. J., Anderson, J. E., & Flin, R. (2012). Non-technical skills (NTS) for enhancing patient safety: Achievements and future directions. *Japanese Journal of Quality and Safety in Healthcare, 7*(4), 360–370.

Koch, R. (1998). *The 80/20 principle*. New York: Currency Doubleday.

Kolcaba, K. (2003). *Comfort theory and practice: A vision for holistic health care and research*. New York: Springer Publishing Company.

Kurtz; C. F., & Snowden, D. J. (2003). The new dynamics of strategy: Sense-making in a complex and complicated world. *IBM Systems Journal, 42*(3), 462.

Lagercrantz, D. (2016). *Syndefall i Wilmslow*. Gyldendal, ISBN 978-82-05-49918-8 (in Norwegian).

Lindegjerdet, F. (2014). *Måløyraidet 1941*. Norway: Ares Forlag AS.

Lindemann, U., & Lorenz, M. (2008). Uncertainty handling in integrated product development. In *International Design Conference—Design 2008, Dubrovnic, Croatia*, May 19–22.

Lipshitz, A. R., & Parsons, S. P. (Eds.). *E-learning: 21st century and challenges*. Hauppauge, NY, USA: Nova Science Publishers.

Magnusson, C. (2001). *Introduktion till designmetodik, rehabiliteringsteknik och design*. Certec, Lund Technical University, Sweden. www.certec.LTH.com (in Swedish).

Malhotra, A., Majchrzak, A., Carman, R., & Lott, V. (2001). Radical innovation without collocation: A case study at Boeing-Rocketdyne. *MIS Quarterly, 25*(2), 229.

Markides, C. C. (2000). *All the right moves*, Harvard Business School Press. ISBN 0-87584-833-8.

Marshall, G. (2005). The purpose, design and administration of a questionnaire for data collection. *Radiography, 11*(2).

McAlpine, H. C., Hicks, B. J., & Culley, S. J. (2008). Investigating the structure and organisation of engineering logbooks for more effective re-use. In *International Design Conference—Design 2008, Dubrovnik, Croatia*, May 19–22.

McDonald, M. (2007). *Marketing plans* (6th ed.). Oxford, England: Butterworth-Heinemann. ISBN 978-0-7506-8386-9.

McMahon, C., et al. (2002). *Opportunities and challenges of knowledge management*. Wuhan, China: TMCE.

McQuarrie, E. (2005). *The market research toolbox: A concise guide for beginners*.

Mehri, D. (2005). *Notes for Toyota-Land: An American engineer in Japan*. London: Cornell University Press.

Meyer, E. (2017). Being the boss in Brussels, Boston, and Beijing. *Harvard Business Review*, July–August 2017 issue.

Montuori, A. (2003). The complexity of improvisation ant the improvisation of complexity: Social science art and creativity. *Human Relations, 56*(2), 237.

Montuori, A., & Donnelly, G. (2018). Transformative leadership. In J. Neal (Ed.), *Handbook of personal and organizational transformation*. Berlin: Springer.

Moore, G. A. (2004). Darwin and the demon. *Harvard Business Review*.

Morris & Hought. (1987). *The anatomy of major projects: A study of the reality of project management*. Oxford: Wiley.

Mulgan, G. (2007). *Ready or not? Taking innovation in the public sector seriously*. Nesta, UK. www.nesta.org.uk/publications.

Nationalencyklopedin. (1994). Malmö, Sweden. www.ne.se.

New, S. J. (2007). Celebrating the enigma: The continuing puzzle of the Toyota Production System. *International Journal of Production Research, 45*(16), 3545–3554.

Nordin-Hultman, E. (1998). Det pedagogiska rummet speglar vår syn på barn. *Pedagogiskt magasin, no., 3,* 22–26. (in Swedish).

Norell, M. (1999). *Managing integrated product development, critical enthusiasm, contributions to design science*. Trondheim/Lyngby: Norwegian Research Council.

Norman, D. A. (1988). *The psychology of everyday things*. New York: Basic Books.

Norlin, A., & Burman, J. (2008). *Här kommer grävskopan*. Rabén & Sjögren. ISBN 978-91-29-66977-0 (in Swedish).

Norrestranders, T. (1999). *The user illusion: Cutting consciousness down to size*. London: Penguin Books Ltd. ISBN 0140230122.

Noyes, K. (2014). *Why investors are pouring millions into crowdfunding*, Fortune, April 17.

Nyström, H. (1996). Företagsledning i ett kreativt perspektiv. In G. Ekvall (Ed.), *Navigatör och inspiratör*. Sweden: Studentlitteratur, Lund. (in Swedish).

OECD. (1997). *The measurement of scientific and technological activities*. Organisation for Economic Co-operation and Development, EU. http://www.oecd.org/science/inno/2367580.pdf.

Olson, J. (2002). Research about office workplace activities important to US businesses—And how to support them. *Journal of Facilities Management, 1*(1), 31–47.

Olsson, F. (1976). *Systematisk konstruktion: en studie med syfte att systematisera innehåll och metoder i samband med produktkonstruktion*. Ph.D. thesis, Lund University, Sweden (in Swedish).

Olsson, F. (1985). *Integrerad Produktutveckling*. Stockholm, Sweden: Mekanförbundet. (in Swedish).

Ordanini, A. (2009). Crowd funding: Customers as investors. *The Wall Street Journal*.

Ottosson, S. (1996). Dynamic product development—Findings from participating action research in a fast new product development process. *Journal of Engineering Design, 7*(2), 151–169.

Ottosson, S. (1997). Planetary organizations. *Technovation—The International Journal of Technological Innovation and Entrepreneurship, 19,* 81–86.

Ottosson, S. (1998). Qualified product concept design needs a proper combination of PAD-MAD before PDM. *Journal of Engineering Design, 9*(2), 107–119.

Ottosson, S. (2003). Participation action research—A key to improved knowledge of management. *Technovation—The International Journal of Technological Innovation and Entrepreneurship, 23,* 87–94.

Ottosson, S. (2004a). Dynamic product development—DPD. *Technovation—The International Journal of Technological Innovation and Entrepreneurship, 24,* 179–186.

Ottosson, S. (2004b). Stage-Gate™ vs. dpd in software development. In *IPD Symposium in Magdeburg, Germany*, September.

Ottosson, S. (2004c). Dealing with innovation push and market need. *Technovation—The International Journal of Technological Innovation and Entrepreneurship, 24*, 279–285.

Ottosson, S. (2006). *Handbook in innovation management—Dynamic business & product development*. Sweden: Tervix AB.

Ottosson, S. (2013). Universal product development. In *International Conference on Engineering Design—ICED'13, Seoul, S. Korea*.

Ottosson, S. (2014). *How to follow up and learn from project work?* Workshop at Gjøvik University College, Norway, October 16.

Ottosson, S., & Björk, E. (2000). Participation action research—A method to improve knowledge of product development and innovation management. In *Tools and Methods in Competitive Engineering (TMCE-2000)*.

Ottosson, S., & Holmdahl, L. (2007). Web based virtual reality. *Journal of Engineering Design, 18* (2), 103–111.

Ottosson, S., Moldavska, A., Ogorodnyk, O., & Skogsrød, T. (2016). What is and how to develop sustainable innovation? *The International Conference on Leadership, Innovation and Entrepreneurship as Driving Forces of the Global Economy (ICLIE), The Palm, Dubai, U. A.E.,* , April 20–22.

Ottosson, S., & Skogsrød, T. (2016). Some experiences from using 3D-printing in the modelling of a new product. In *NordDesign 2016, Trondheim*, August.

Pagnini, F., & Phillips, D. (2015). Being mindful about mindfulness. *The Lancet Psychiatry, 2*(4), 288–289. https://doi.org/10.1016/S2215-0366(15)00041-3.

Pech, R. J. (2001). Reflections: Termites, group behavior, and the loss of innovation. *Journal of Managerial Psychology, 16*(7/8), 559.

Pinker, S. (2007). *The stuff of thought*. London: Penguin Books.

PMBOK. (2000). *A guide to the project management body of knowledge* (PMBOK Guide 2000 Edition). Project Management Institute. ISBN 1-880410-81-8.

Pressman, R. S. (2000). *Software engineering (A practitioner's approach)* (5th ed.). New York: Mc Graw-Hill Education.

Pugh, S. (1991). *Total design: Integrated methods for successful product engineering*. New York: Addison-Wesley. ISBN 0-201-41639-5.

Quinn, D. (1985). *Rewriting the corporate brain*. New York: Berret-Kohler Publishers.

Rama, Y., & Herbig, P. (1996). Global markets and the new product development process. *The Journal of Product and Brand Management, 5*(6), 38.

Ramage, J. D., & Bean, J. C. (1998). *Writing arguments* (4th ed., pp. 81–82). Needham Heights, MA: Allyn & Bacon.

Rasulzada, F. (2007). *Organizational creativity and psychological well-being*. Ph.D. thesis, Department of psychology, Lund University, Lund, Sweden.

Renton, J. (2009). Coaching and mentoring: What they are and how to make the most of them. New York: Bloomberg Press. ISBN9781576603307.

Robert, J. B. (2015). *Holacracy—The revolutionary management system that abolishes hierarchy*. London: Penguin Random House. ISBN978-0-241-20585-3.

Robertson, B. J. (2015). *Holacracy: The new management system for a rapidly changing world*. New York: Henry Holt and Company.

Romme, A. G. L., & Endenburg, G. (2006). Construction principles and design rules in the case of circular design. *Organization science, 17*(2), 287–297.

Romme, G. (2016). *The quest for professionalism: The case of management and entrepreneurship*. Oxford: Oxford University Press. ISBN: 9780198737735.

Rosenfeld, R. (2006). *Making the invisible visible: The human principles for sustaining innovation*. Bloomington, IN: Xlibris Corportation.

Rogers, E. M. (2003). *Diffusion of innovations* (5th ed.). New York: Free Press.

Rowland, G. (2004). Designing with: A homeopoietic ethic for organizational change. *Systems Research and Behavioral Science, 21*(3), 219.

Sample, J. (2004). The Myers-Briggs type indicator and OD: Implication for practice from research. *Organization Development Journal, 22*(1), 67.

Sarasvathy, S. D. (2001): *What makes entrepreneurs entrepreneurial?* Submitted to Harvard Business Review.

Sarasvathy, S. D. (2006a). *New venture performance*. Carlottesville, VA, USA: University of Virginia Darden School Foundation.

Sarasvathy, S. D. (2006b). *The entrepreneurial method: How expert entrepreneurs create new markets*. University of Virginia Darden School Foundation, Carlottesville, VA, USA.

Schmidt, T. S., Böhmer, A. I., Wallisch, A., Patezold, K., & Lindemann, U. (2017). Media richness theory in agile development. In *23rd Conference on Engineering, Technology and Innovation (ICE/TMC), Madeira, Portugal*, June 27–29.

Schmidt, T. S., & Patezold, K. (2016). *Agilität als Alterative zu traditionellen Standards in der Etwicklung physischer Produkte: Chancen und Herausforderungen*. In *DfX-Symposium 2016, Jesteburg, Germany*, October 05–06 (in German).

Schneider, B. (2000). Managers as evaluators: Invoking objectivity to achieve objectives. *The Journal of Applied Behavioral Science, 36*(2), 159–174.

Schrage, M. (2000): *Serious play—How the world's best companies simulate to innovate*. Harvard Business School Press.

Schumpeter, J. A. (1934). *The theory of economic development*. Cambridge, Massachusetts, USA: Harvard University Press.

Shkliarevsky, G. (2017). Understanding the process of creation: A new approach. *Journal of Sustainable Business and Management Solutions in Emerging Economies, 22*(3).

Siau, K. L., & Yang, Y. (2017). Impact of artificial intelligence, robotics, and machine learning on sales and marketing. In *MWAIS 2017 Proceedings*. http://aisel.aisnet.org/mwais2017/48.

Smith, P. G., & Reinertsen, D. G. (1992). Shortening the product development cycle. *Research-Technology Management, 35*(3), 44–49.

Starcevich, M. M. (2016). *Coach, mentor: Is there a difference?* http://www.coachingandmentoring.com/Articles/mentoring.html.

Story, M. F., Mueller, J. L., & Mace, R. L. (2001). *The universal design file: Designing for people of all ages and abilities*. Raleigh, North Carolina: North Carolina State University.

Stump, R. L., Athaide, G. A., & Joshi, A. W. (2003). Managing seller-buyer new product development relationships for customized products: a contingency model based on transaction cost analysis and empirical test. *The Journal of Product Innovation Management, 19*(6), 439–454.

Swan, H., & Furuhjelm, J. (2010). Creating value through lean product development—Applying lean principles. In *NordDesign 2010, Göteborg, Sweden*, August 25–27, 2010 (pp. 489–497).

Tabrizi, B., & Walleight, R. (1997). Defining next-generation products: An inside look. *Harvard Business Review*, 116–124.

Tidd, J., Bessant, J., & Pavitt, K. (2001). *Managing innovation*. Chichester: Wiley.

Torkildsby, A. B. (2014). *Existential design—Revisiting the "dark side" of design thinking*. Ph.D. thesis at University of Borås, Sweden.

Tseng, M. M., & Jiao, J. (2001). *Mass customization in: handbook of industrial engineering, technology and operation management* (3rd ed., p. 685). New York: Wiley. ISBN 0-471-33057-4.

Tuckman, B. W. (1965). Developmental sequence in small groups. *Psychological Bulletin, 63*.

U.S. Bureau of Census. (1997). *1996 survey of income and program participation*. http://www.census.gov, August–November.

Ullman, D. G. (2003). *The mechanical design process*. New York: McGraw-Hill.

Upitis, R. (2004). School architecture and complexity. *Complexity and Education, 1*(1).

Utterback, J. M. (1994). *Mastering the dynamics of innovation*. Boston: Harvard Business School Press.

von Hippel, E. (2005). *Democratizing innovation*. Cambridge, MA: The MIT Press. ISBN 0-262-00274-4.

Ward, A. C. (2007). *Lean product and process development*. Cambridge, MA, USA: The Lean Enterprise Institute.

Westergård, R. (2016). *Ett jordklot räcker*. Stockholm, Sweden: Calidris Förlag. ISBN 978-91-983339-3-0 (in Swedish).

Wetter-Edman, K. (2014). *Design for service. A framework for articulating designers' contribution as interpreter of users' experience*. University of Gothenburg, Doctoral dissertation.

Wheatly, M., & Kellner-Rogers, M. (1999). *A simpler way*. San Fransico: Berrett-Koehler. ISBN: 1-57675-050-7.

Wikström, L. (2002): *Produkters Budskap, Metoder för värdering av produkters semantiska funktioner ur att användarperspektiv (in Swedish)*. PhD thesis at Department of Product and Production Development at Chalmers University of Technology, Gothenburg, Sweden.

Wolf, F. A. (1989). *Taking the quantum leap—The new physics for non-scientists*. New York: Harper & Row Publ.

Womack, J. P., Jones, D. T., & Roos, D. (1990). *The machine that changed the world*. New York: Simon & Schuster.

Yan, H. S., & Jiang, J. (1999). Agile concurrent engineering. *Integrated Manufacturing Systems, 10*(2), 103.

Yapps-Cohen, L. (2017). www.amcreativityassoc.org/Resources/Documents/Cohen-TOP+10+Reasons+Why+We+Need+INNOVATION.pdf.

Index

S. Ottosson, *Developing and Managing Innovation in a Fast Changing and Complex World*, https://doi.org/10.1007/978-3-319-94045-8

T

MIX
Papier aus verantwortungsvollen Quellen
Paper from responsible sources
FSC® C105338

If you have any concerns about our products,
you can contact us on
ProductSafety@springernature.com

In case Publisher is established outside the EU,
the EU authorized representative is:
Springer Nature Customer Service Center GmbH
Europaplatz 3, 69115 Heidelberg, Germany

Printed by Libri Plureos GmbH
in Hamburg, Germany